THE NEW FOOD CHAIN:
An Organic Link
Between Farm and City

THE NEW FOOD CHAIN: An Organic Link Between Farm and City

Edited by Jerome Goldstein
of Organic Gardening and Farming

Rodale Press, Inc.
Book Division
Emmaus, Pa. 18049

Standard Book Number
0-87857-050-0

Library of Congress
Card Number 72-88823

Grateful acknowledgment is made for permission to reprint in this book:

A New Policy Direction for American Agriculture by Marion Clawson, Director, Land Use and Management Program, Resources for the Future, Inc. from The Journal of Soil and Water Conservation.

Where Cities and Farms Come Together copyright© by Wendell Berry, an address for the most part extracted from the essay "Discipline and Hope" in his volume, A CONTINUOUS HARMONY. Reprinted by permission of Harcourt Brace Jovanovich, Inc.

The Importance of Cities to Rural Living. Abridged from "The Economy of Cities" by Jane Jacobs. Copyright© 1969 by Jane Jacobs. Reprinted by permission of Random House, Inc.

Printed in the U.S.A.
on recycled paper
JB-1

First Printing—February, 1973

CONTRIBUTORS

Ruth C. Adams is one of the nation's leading writers on citizen activism, developing the premise that individuals and groups hold the key to saving the environment. Her books include *Say No!* and *Did You Ever See a Fat Squirrel?*

Floyd Allen is west coast editor of *Organic Gardening and Farming* and has written forcefully and sensitively about the common aspirations of farmers and consumers.

Peter Barnes is the west coast editor of *New Republic,* and a co-founder of the National Coalition for Land Reform.

Wendell Berry is professor of English at the University of Kentucky (named Distinguished Professor in 1972), novelist, poet, and organic farmer. His published works include *Farming: A Handbook, The Hidden Wound, The Long-legged House, Nathan Coulter, A Place on Earth, Openings,* and *Findings.*

Marion Clawson, an economist, directs studies in land use and management for Resources for the Future in Washington, D.C.

Wayne H. Davis is a professor in the School of Biological Sciences at the University of Kentucky and is well-recognized for his analyses of environmental problems.

Jerome Goldstein is executive editor of *Organic Gardening and Farming,* editor of *Environment Action Bulletin,* and author of *How to Manage Your Company Ecologically* and *Garbage As You Like It: A Plan to Stop Pollution by Using Our Nation's Wastes.*

Jane Jacobs is the author of *The Economy of Cities* and *The Death and Life of Great American Cities,* and has served as editor of *Architectural Forum* from 1952 to 1962.

Gene Logsdon is associate editor of *Farm Journal* magazine and the author of *The Wyeth People* and *Two Acre Eden*

Robert Rodale is the editor and publisher of *Organic Gardening and Farming, Prevention,* and *Fitness for Living;* the syndicated columnist of *Organic Living;* and the author of *Sane Living in a Mad World.*

CONTENTS

INTRODUCTION

A major obstacle to solving some of society's present problems can be traced to the barriers separating people who live in cities and people who live on farms. This collection of ideas, essays and ramblings is one attempt to show that such barriers have no basis for existence. Once the walls separating cities and farms crumble, we can get on to the serious business of improving life in each area. For too many years, our elected officials in Washington, D.C. as well as in state capitals, have reacted to urban and rural decay as if it were an either-or phenomenon. And sadly, the quality of life in city and farm deteriorates simultaneously. There is no escape route for one at the expense of the other. We must stop using city people as a scapegoat for farm problems, and rural people as a scapegoat for urban problems. Once the link is made clear, some effective government programs can be developed. Instead of having representatives of city voters speak against farm-oriented programs, and vice versa, we will be seeing the adoption of programs and alternatives which stress the common benefits to both city and rural people.

Perhaps the most eloquent description of the link between city and farm is provided by Wendell Berry, poet-professor at the University of Kentucky, and an organic farmer. His "Where Cities and Farms Come Together" concludes this book. But national goals and life quality are now becoming, fortunately, intertwined with much simpler concepts—like food quality. A recent survey revealed that 36 per cent of Americans said that there are too many food additives and food is no longer safe. More than one-third of our nation are concerned about the safety of their foods and are more conscious of good nutrition. More than one-third of the voters in this country are asking for ways to grow better food on the farm. More than 70 million Americans care enough about the food chain to be dissatisfied with the status quo—agri-business methods that have perverted quality of food and life, even if the motives supposedly were the best and most food at the cheapest price.

The net result is almost exactly the opposite. *Hard Tomatoes, Hard Times,* a study of the land grant college complex, reveals how much of our tax dollars wind up serving "an elite of private, corporate interests in rural America, while ignoring those who have the most urgent needs and the most legitimate claims for assistance."

A hope, a vision, a dissatisfaction, an understanding—these forces are combining to form *The New Food Chain: An Organic Link Between Farm and City.*

Organic Force

Jerome Goldstein

The organic idea is basically a simple idea. It only takes a very few words to describe what organic means in the garden . . . or what organic means on the farm. The explanation can take less than a minute—begin with the soil, get into the compost heap, the natural cycle, the need to return garbage and sludge and wastes back to the land, the hazards of pesticides and artificial fertilizers to the environment, and the personal health benefits that go with eating quality, nutritious food.

Not only is the organic idea basically a simple one, but it also is a personal one. Organic gardeners and farmers have developed countless ways to accomplish what they set out to do—different planting dates, planting methods, varieties, equipment, tilling techniques, composting methods —and on and on. Organic gardeners and farmers are a most individualistic group. *Organic Gardening and Farming,* as a magazine, is shaped far more by its readers than by its editors.

Yet more and more in the past few years, some grandiose concepts have been getting mixed in with the compost heap. Though it's still the same old compost heap as it always was, in the 1970's the heap has come to mean something more. It still is simple. It still is personal. But now, instead of only rating that heap for its ability to make a soil fertile, we talk about it in such terms as social practice and harmony with the environment. To many people, it's still just a pile of garbage and manure—true, a controlled pile on its way to becoming humus—but nevertheless, just a pile. To others, it's a vision of a society in harmony with the environment.

Yet the visionary qualities of the organic idea are only valid as long as the compost heap exists and what's more heats up. The beauty of the organic idea as a social force is that it is firmly rooted in materials, methods and efforts which most people usually refer to as garbage, idealistic, inefficient or economically unfeasible. And as more and more people see it, the organic idea provides a model route from where we are now to where we would like to be in the future. "Organic force" has become meaningful in the marketplace and on the farm, in the supermarket and in the classroom because it has become the codeword for something even more significant than no pesticides, chemicals or additives. The word organic is becoming a linking symbol upon which a consumer can relate to a producer. It is a substitute for national brand advertising via television, newspapers or magazine; the word organic when truly defined cannot have a national brand name because its essence is its localization and personalization.

The organic concept has the ability to show government officials, industry executives, professors and policy-makers everywhere how smallness can become economically viable —how money can be spent to subsidize projects other than those leading to mechanization, bigness, and so-called efficiency.

The organic concept is a forerunner of how an economic base can be given to idealistic concepts. For large-scale development, it will be necessary for new laws to be passed; it will be necessary for new governmental policies to go into effect. But even now, organic force is developing models for the survival of mankind. In less grandiose terms, organic force is showing how some people can make a buck and spend a buck while accomplishing some good things. It shows how ecology can be blended into daily living without doing anything special. It offers us a game plan—a personal action plan—that takes us beyond wringing our hands, preaching and so forth. This organic force may very well be our best reason to be optimistic at this time.

Obstacles to the survival of mankind have been written about and *theorized about* by many eminent scholars with a variety of scientific, economic and social expertise. One of the clearest pictures of where we should be heading is contained in a document called the "Blueprint for Survival" developed by *The Ecologist,* a British publication edited by Edward Goldsmith.

2

To succeed, says the "Blueprint for Survival," we must formulate "a new philosophy of life, whose goals can be achieved without destroying the environment, and a precise and comprehensive program for bringing about the sort of society in which it can be implemented."

Mr. Goldsmith, his fellow editors of *The Ecologist,* and the scientists who contributed to "Blueprint," point out that we behave as if we knew nothing of the environment and had no conception of its predictability. Man's aim must become "a stable society" characterized by a steady or declining population, decentralized living and strict limits on using resources. How else can we adjust as a civilization to the fact that indefinite growth of whatever type cannot be sustained by finite resources.

In developing a strategy for change to a stable society, "Blueprint" gets into concepts which have been continually discussed and developed by organic gardeners and organic farmers in this country—minimum disruption of ecological processes; maximum conservation of materials and energy; a social system in which the individual can enjoy, rather than feel constricted by those conditions; systematic substitution of the most dangerous components of present technology with ones that cause minimum disturbance to normal processes of the ecosphere; decentralization of economy at all levels, and the formation of communities small enough to be reasonably self-regulating and self-supporting; and education for such communities.

Based on the information we have received as editors of *Organic Gardening and Farming* magazine, we believe—as do the creators of "Blueprint for Survival"—that it is possible to change from an expansionist, over-chemicalized, over-plasticized, over-additived, over-dumping society, to a stable, more natural, more personal, more organic society without loss of jobs, with less starvation and malnutrition than we are now plagued with, and without an increase in real expenditure. In discussing the real-life implications of these concepts, one immediately must think of how they relate to food production—land use, fertilizer use, pesticide use, corporate farming vs. family farming, food distribution, marketing and not incidentally, disposal of food when it becomes garbage and sewage sludge.

And these considerations lead right back to the backyard garden, the small farm, the compost heap, the organic foods, the mama-and-papa neighborhood grocery store. All have become the symbols of the new American Revolution.

Social Significance of Organic Foods

A whole generation of Americans has grown up without any personal communication with the producer of their foods. The supermarket check-out clerk has been the closest human contact, as the food goes from shelf and freezer into shopping basket for transfer to closet and freezer. The television screen gives the clearest picture of where the food came from before it got to the supermarket.

Organic foods have the ability to turn this all around. The consumer can identify the farmer, and the farmer can identify the consumer. The human identity of each can surface and interrelate. The money spent for food—for its production and consumption—can become a real economic force for societal and environmental objectives.

Part of each dollar spent for organically-grown foods should mean:

1. Less money spent on pesticides and artificial fertilizers—consequently, less production of chemical pollutants and contamination of the environment.

2. More money for the farmers and farm workers who make their homes on the land—thus more profits when land is used for producing food instead of for residential and industrial development.

3. More jobs with adequate compensation on farms growing crops by labor-intensive organic methods—thus less forced migration to city ghettos.

4. More economic incentive to bring composted organic wastes from the city onto farmland where it builds up humus content.

5. More economic support to the *small* entrepreneur— the family farmer, the mama-and-papa grocery store, and the *local* brand name.

4

6. More demand for personal services and less for mass-distributed environmentally-hazardous products. That means more money for the man who is educated to advise on how to recycle wastes back to the land and less for the chemical fertilizer salesman whose product is supposed to be a cure-all for everything (except the pollution it brings to waterways from runoff.)

7. More stress to have local farmers supply the local market.

When J. I. Rodale started a small magazine with a handful of subscribers back in 1942, neither he nor anyone else could predict its role 30 years later. In fact, even today, only a handful are still aware of how organic force relates to the very core of many environmental solutions.

Yet today, the simple ideas have taken on a new aura of social significance. Organic methods in agriculture put an economic base under city planners' dreams of open spaces around urban areas. For city people to get high-quality, inexpensive organic foods, city people need country people on small farms. A good many of those farms should be close to the city. Every city in America faces a crisis with garbage disposal; local farms know that garbage can be useful once it's returned safely to the soil. The organic idea lets everyone understand how everything interrelates.

Of course, if organic foods are to have all the relevance implied above, tremendous demands are put upon the buyers of those foods. Consumers constantly must ask (often more than once) where does the food come from—whose farm, what growing methods, who knows the farmer—and have you given him a chance to know you?

If the food is expensive, you should find out why. Has the grower's harvest been drastically reduced by insects or disease? Or is it that the food has come several thousand miles, or, one or more links in the food distribution chain are taking excess advantage of your desire for organic foods?

When you buy organic foods, you should be aware that you are using your dollar to encourage change in American agricultural methods. For years, American farmers have

5

been led to believe that food is no different than any other product. Churn it out in assembly-line fashion as fast and as cheaply and mechanically as you can. Just like we do with cars. Or envelopes. Or wigs. Or any other product.

But now you are taking your food dollar and you are dissenting. You are saying that all food is not equal in quality regardless of how it is grown.

By your dissent, by your desire to buy organic foods, you are leading a consumer food revolution. The food revolution begins with your buying organically-grown foods. Its ideology is even good for your stomach.

When you buy organically-grown foods produced by a family farmer who is not supposed to be able to make a living on the land, you become an organic force helping to reverse a trend that has driven people off the land and made farming an old man's profession. By buying organic foods at a mama-and-papa neighborhood store that is not supposed to be able to compete with supermarket chains, you are helping to change the make-up of America.

I believe millions of Americans are saying important things when they buy organic foods. Things like "land ethic" and "decent wages for more farm workers" and "a way for the little guy to compete" and "I don't want everything to look alike—especially my food."

We are only beginning to understand how much force—how much political and economic clout—can be generated by the rapid expansion of the organic market. But that clout will only develop if the market develops for the reasons cited. If that market develops a corporate structure of its own, then the societal and environmental benefits diminish—even if the quality of the food harvested is maintained.

Organic food buyers have the potential to keep family farms economically healthy. The term "organic" becomes a brand identity which normally costs millions of dollars in corporate advertising to establish and maintain—and by its nature, excludes the big spenders, and is available only to the family farm.

What we are trying to show is that there is a rapidly growing trend for food grown with more man-power. Sure, we know that eggs produced by chickens which are allowed to run around will cost more than eggs produced by chickens that are penned four to a cage. But those organically-produced eggs will get and deserve a higher price. Part of that higher price will provide decent wages for more farm workers—right now those farm workers of the future may be sweating it out in some urban tenement.

Does that mean organic foods are always higher-priced and only for the upper middle class? We think not—but time will tell. Many environmentalists including organic gardeners have been worried that industry is trying to isolate them from labor and the poor. Statements that organic foods are only for the rich, that anti-pollution costs force layoffs and claim funds needed to eliminate poverty are seen as an effort to set workers and the poor against environmentalists.

Groups like the Rural Advancement Fund have shown the relevance of the organic idea to small farmers and rural poor, and the vital necessity to use organic foods as a way to relate the farmer to the consumer. And to relate what is wrong in the city with what is wrong on the farm. Organic foods can help to solidify the coalition between the needs of sharecroppers and small farmers and the environmentalists' concern for the right land use.

A Rebirth in the Ability to Communicate

It is most important to dwell on the food aspect of the organic force, since it is so tangible. Food is the great link between problems in the city and problems on the farm. Therefore food becomes the great force to communicate between the people who consume the foods and the people who grow the foods. Bigness in food production methods and marketing has come to mean anonymity—and a breakdown in communications. Smallness in food production and marketing has come to mean personal identity—and a rebirth in communications.

The great strength in organic force is how it has led to communication between individuals and groups who would not recognize each other if it were not for their common interest in organic foods. Someone remarked the other day that he thought a great many of the readers of *Organic Gardening and Farming* religiously watch Lawrence Welk. That's true. A great many listen to the Beatles also. A great many are right-wing Republicans, and a great many are what used to be referred to as left-wingers. But organic force has a way of pointing out the shallowness of those labels when compared to the basic values people share with each other.

There are those social commentators who point to the divisive effect of environmentalists. They say environmentalists have focused attention on fish and wildlife, and away from people . . . that the ecology movement is a seductive diversion from the political tasks of our time . . . that it is only the disillusionment of the left which joins the longer-standing isolationism of the right and mixes with the indifferent middle. Such views place organic force as still another drive against the poor, and that a quality environment is meant only for those who can afford it. According to these critics, the environmental crisis is largely the result of doing too much of the *right* sort of thing. For example we are told that we have "succeeded in cutting down the mortality of infants, succeeded in raising farm output sufficiently to prevent mass famine, succeeded in getting people out of tenements and into the greenery and privacy of single-family suburban homes. "Now, say these observers, "we must build on that success and determine how to trade off between a cleaner environment and unemployment . . . harmonize the worldwide needs of the environment with the political and economic needs of other countries, prevent the environmental crusade from becoming a war of the rich against the poor."

These critics of environmentalists—no matter how noble their motives—are practicing a brand of eco-pornography that is both subtler and more dangerous than the oil or electric companies have ever done. At a time when divisive walls of old ideologies are falling down and new vital alliances can be built, those critics should think a little harder and a little longer before declaring themselves as the only ones defending people.

We are going through a rather special time right now when dissatisfaction and unrest are so widely evident that each of us—conscious of such a phenomenon as organic force—wants to do everything we can to translate that dissatisfaction into something positive—something that will truly improve the quality of life while improving the quality of the environment. And a major part of the effort must be to show how widespread is this organic force.

Organic force comes from the black sharecropper in Georgia, and from the volunteer in a Park Avenue office for consumer action, and from an ecology center in Washington, D.C.; and from a dormitory group in Ann Arbor; and from a senator's office on the Hill; and from a community change center in an urban ghetto; and from millions and millions of poor and rich alike, health food customers and supermarket buyers—many using food stamps —farmers and migrant laborers.

As organic force gathers strength and becomes more obviously a factor for societal change, it must be made clear that there is no single spokesman or center or headquarters from which these forces spring forth. If anything is really going to get done, it must be clear to all that organic force is spelled with a lower case "o" and "f". Readers of *Organic Gardening and Farming,* I believe, have been among the leaders of these forces—but the real strength is their action as individuals not as subscribers to a magazine. Just as no magazine can harness these forces, so no single political party, economic group or any other organization can become its master.

New Jobs, New Profits in "Mama-and-Papa" Stores, Farms, Services

About once a year, for the last 3 years, several of us decide to make a pilgrimage to the Department of Agriculture in Washington, D.C. to establish a more direct contact between the organic farming community and the USDA. This year's expedition took place a few weeks ago.

I had decided to ask the USDA to name an "organic" man within the Department—someone who could become intimately aware of the needs of an organic, family farmer and try to relate what was going on within the Department to those needs. To make my point, I told the USDA official that there were now more than 800,000 subscribers to *Organic Gardening and Farming* magazine and it was fair to assume that in that number there were between 10,000 and 50,000 farmers who could benefit from USDA recognition of their existence.

I did not get a positive response to my request, since the official pointed out that the USDA served all farmers, but the official was most impressed with the figure 800,000. "How," he asked, "could *Organic Gardening* magazine have 800,000 subscribers and be thriving, and *Look* Magazine be out of business?"

And the answer to that question, I maintain, is an indication why the organic force can supply an economic base to far more than a publishing company and its employees in Emmaus, Pennsylvania.

I have no predictions to make about the future of mass marketing, but I do feel certain that one-to-one personalized and specialized marketing, so basic to an organic force, holds the key to jobs and profits for a great many Americans in the future.

We are learning this lesson too often at great hardship. The day before Christmas, 1971, the *Wall Street Journal* described how large retail department stores were probably the biggest beneficiaries of the 6% unemployment rate. Engineers, computer specialists, schoolteachers, technicians and recent college graduates flocked to the stores last Christmas season in the hopes of landing a temporary job during the rush. The quality of the temporary help was the best in 6 years. But not all stores were happy. Some workers spent "an inordinate amount of time in informal conversation with customers" instead of hurrying on to the next sale.

That anecdote tells a great deal, I believe. Unemployed technicians, working as clerks, treating customers as human beings—and they're out of step in the mass market department store. Treating people like human beings is not a true cost of product distribution.

There is a fantastic need for an alternative in the marketplace, every marketplace, not just in supermarkets and food-growing and distribution. People ask us why there is such a surge in interest in organic foods today. In response, someone will answer "ecology." But ecology unfortunately is an impersonal concept—an oh-so-big concept that creates a picture of one earth when what we really seek is a vision of one person, and then a second person, and a third.

Organic foods and organic force are so powerful because first there is a personal picture—first it's me—then it's the farmer—then it's the store owner (he knows my name, and I know his)—and then it gradually builds to include everything and everyone on our planet. But first it was personal! The organic force can bring people back into the marketplace, and pay for them to be there. It will take tremendous re-shaping of federal subsidies and government spending to enable that organic force to employ great numbers of people. But here again, without subsidies or government aid, organic force is leading to modest commercial success for many innovative individuals.

To attack the employment crisis of the 1970's we've got to shift gears radically so that we no longer equate assistance with machines, highways, and super-technology. We need to develop and employ highly-trained individuals who take the adversary role—who ask the tough questions before research and policies and new product development decisions are made. These individuals must play a role in education as in industry, in government as in research. Not only will this approach provide jobs, it will also mean that most highly-spirited, highly-motivated men and women can work within the system to improve our society instead of being driven out.

Writing in the *New Scientist,* the Honorable Reginald Prentice, who had been Britain's Minister of Overseas Development from 1967 to 1969, speculates on the inadequacy of conventional approaches to get 280 million more jobs. He sees the need for drastic change in three different areas: a shift to labor-intensive projects; greater emphasis on the rural sector of the economy; a shift of emphasis in the education systems so that they are more closely geared to employment opportunities. What we refer to as organic force must show up in each area mentioned by Mr. Prentice.

"Every project should be assessed primarily in terms of the effect on employment. This will mean a movement away from sophisticated methods towards intermediate technology. It will mean a larger number of smaller projects and fewer prestige projects. . . . Urban problems can only be tackled successfully against a background of rural development. . . . It also means a much more urgent need for land reform in many countries—otherwise there is a danger that only the big farmers and land owners will benefit from these trends and that they may use mechanized methods which lead to further unemployment. The educational system has to move rapidly towards vocational training, especially for middle level jobs, and especially so as to offer training that is dovetailed to the needs of the rural areas."

An economics professor at Roosevelt University in Chicago, Walter Weisskopf, has also done an excellent job of relating GNP fetishism to individual strangulation. He develops his views in a book called *Alienation and Economics*. "There are human potentialities and needs which are not related to the procurement of goods and services and cannot be satisfied by producing and buying and selling." Professor Weisskopf lists these as: "Love, friendship; primary, warm, affectionate human relations." (And remember those talkative college-educated clerks annoying the hell out of the department store managers.) For Professor Weisskopf, for alienation to be overcome, "revolution and reform have to start with the individual and from within." Again, we are reminded to think small and personal if we are to solve the grand problems of our age.

More and more, as I speak to people concerned with improving the environment, I find that genuinely successful efforts that are both environmentally-and-economically sound occur when the individuals involved THINK PERSONAL AND SMALL! When a project (like recycling newspapers) involves one town as opposed to an entire state, or one type of product or service as opposed to an entire range of products or services, repeatedly, I find rea-

12

sons for optimism when the product or service is developed in terms of the personal "mama-and-papa" approach and only reasons for pessimism when the development is characterized by computer-like super-conglomerate philosophy.

To return to agriculture, what seems to be increasingly evident now is that organic farming methods with their emphasis on nonmonoculture farming, return of organic wastes to the soil, prohibition of pesticides and high-nitrate fertilizers, and labor-intensive agriculture, offer the environment and the farmer a most worthwhile and potentially profitable alternative.

Now, let's take a look at the economics of organic farming. Because it is labor intensive, labor costs go up. Normally, this is bad for the economics, but now we have a product (food) that commands a higher price *because of the hand labor involved.* The handicap thus becomes an asset since more jobs are created in the production of the food.

Organic farming means that pesticides are not used; but pesticides are used in conventional farming to prevent crop damage even though most people consider them environmentally hazardous. What does an organic farmer use to control pest damage? Many rely on biological controls—a relatively unresearched area in the field although there has been some excellent laboratory work. Some entomologists who have done research into biological controls have started their own consulting firms to help organic farmers control pests without pesticides.

Here again, we can see the transfer of costs from an environmentally-hazardous substance like pesticides into an environmentally-sound pest control *service.* Even if the costs are the same, the benefits of an environment more free from pesticides are obvious. And a pest control service providing income and jobs for humans replaces a pest control product which required energy and material from our eco-system to make.

Organic farming, whenever possible, substitutes the personal service for the impersonal product. No longer need there be spray schedules to be rigidly followed with the only beneficiary being the spray maker. Agricultural Experiment Stations in the United States must aggressively address themselves to the needs of family farmers and give advice they need.

Organic farming methods also forge the link between city and farm, since—to farm organically—you need to return organic wastes to the land. It has been suggested that farmers receive payments based on the amounts of organic wastes allowed to be brought to their lands. This subsidy makes great sense. We are already spending great sums of money to burn our city wastes, so it seems eminently economical to direct those dollars and the wastes to the farms. And a subsidy like this certainly makes more sense than damming rivers out west to provide water for corporations to grow grain to glut the market still further.

To get specific, organic farming can be a prime force in building up the humus content of our soils instead of our waterways. Organic farmers—whether or not they benefit directy from subsidies for taking wastes—can get paid for the extra costs of putting composted wastes on their land by getting higher prices from the consumers who want their food grown by such methods. What's more, the standards of organic farming can help keep industrial contaminants from ever getting into the municipal waste stream.

The family farm of today can best be characterized by the one family supported by its operation—by the "mama-and-papa" character it has. The counterpart of the family farm in urban areas is the family store—and here again is the "mama-and-papa" image.

And the same economic ray of hope offered to organic farmers applies to operators of natural food stores.

The "mama-and-papa" foods store is the ideal way to market those kinds of foods we refer to as organically-grown. I believe this for a variety of reasons, some environmental, some economical, and some for a vague feeling that the "natural food concept" is a small idea better suited to personal marketing methods. Here we get into certification standards—the need to confirm the methods used to grow the food on the farm and more important, to know which farm produced the food.

14

Robert Rodale, editor of *Organic Gardening and Farming,* has pointed out that "while the organic method is well-shaped as an idea, it remains to be fleshed out as a technology. What makes conventional farm technology so powerful is its input of science. Vast sums have been spent on scientific research that fed muscle directly into the chemical-oriented, high-power-using, agribusiness technology. Without that scientific input, farming of large areas with little use of manpower would hardly be practical.

"By contrast, organic farm technology has been able to feed only on the scraps and remnants of conventional science. Farming with natural materials, and without poisons, has been considered old-fashioned by most scientific institutions, and hardly any effort has been expended on perfecting ways for families to make small farms more productive within the ecological, organic framework. Although organic farmers can use some parts of conventional farm technology (modern tractors, combines, manure-spreading equipment, and vehicles, for example), the general thrust of agribusiness has been away from naturalness in farming. Therefore, as conventional farming became more mechanized and chemicalized, farmers inclined toward natural methods found themselves at more of a disadvantage.

"What would happen if significant scientific efforts were directed toward creating a more effective organic farm technology? What if the small farmer gradually found himself offered machines, fertilizers, plants, and techniques that would enable him to produce marketable amounts of fresh foods, at a reasonable cost and without the chemicals that are becoming more irritating to food buyers? And what if the creation of those techniques was combined with restrictions on tax and subsidy benefits now enjoyed by large-scale, agribusiness farmers? The answer to those questions is that a farm revolution would happen, accelerated by the growing consumer demand for organically grown food. If small-scale farming without chemicals suddenly were made even more competitive and satisfying to both farmer and consumer than agribusiness and its plastic food, then population of rural areas would stabilize, if not begin to increase.

"Improved small-farm technology is feasible. No tremendous breakthroughs are needed, only the fleshing-out of concepts which have already been outlined. Organic farming, therefore, can be made much more practical and competitive than it now is, with only moderate help in the way of improved science."

The Potentials of Recycling and Composting

The waste disposal business is the third largest area of government expense—following road-building and education. Most of the money goes for collection and transportation to the dump.

Recycling wastes—which all agree is the route to take—needs development. It needs consultants, and all kinds of hardware. It needs special transportation equipment to convey wastes from where they are pollutants to where they can be used. It needs a mass transit system designed for wastes. It needs heavy-duty grinding equipment, and sophisticated sorters and classifiers. It may involve separators based on cryogenics; it may involve the same technology that now is used in oil refineries; research projects indicate that equipment developed in one industry can be adapted for use in the waste recycling industry. It needs land around urban areas on which to apply composted wastes.

Two days after Thanksgiving, the man who is often referred to in newspaper headlines as the "nation's environmental chief" issued a public statement that composting as a means of disposal is economically unfeasible. The man was William D. Ruckelshaus, administrator of the U. S. Environmental Protection Agency (EPA), and recently he has been sounding more and more like a Secretary of Commerce instead of our "environmental chief." We wonder what happened to the garbage from his Thanksgiving dinner. Unless he has a backyard compost pile, it's probably smoldering in a poorly-run landfill . . . floating downstream toward the Atlantic . . . or in minute particles after leaving the incinerator stack.

Eloquent statements on recycling organic wastes have come out of Washington, so we must assume they have something else in mind besides dumping, burning or burying. Research and world-wide operating experience show that the compost process offers a real potential to return some wastes back to the land in an economical, safe and environmentally-sound way. A small percentage of all the compost can be marketed as an income-producing soil conditioner, but the vast amounts must be used to improve our soils instead of destroying our waters.

Unfortunately, the EPA evaluates composting only as a commercial venture in two recent reports: "American Composting Concepts" and "Composting of Municipal Wastes in the United States." At a time when so much talk is given to recycling, we must develop a national policy to return organic wastes to the land. The composting process can help achieve this goal—without a lot of fancy buildings or equipment. But composting—like any recycling system if it's going to succeed—needs a commitment from our elected and appointed leaders. Any improvement in waste treatment will cost us money, but composting is the only process that gets wastes back where they belong!

More money must be spent on recycling. More profits must be made by firms who serve the recycling field. Sooner or later, there will be a guaranteed market for reclaimed wastes. When that happens, the recycling boom will be truly on. Here are a few examples of how treated organic wastes and waste-water are being recycled back to the land:

Los Angeles County: All of the digested sewage solids—about 100 tons of dried solids per day—are being composted in windrows for about 16 days before being turned over to a private firm for marketing as a soil conditioner-fertilizer.

San Francisco: A San Francisco environmental management firm, headed by the highly regarded public health authority Frank Stead, recommends that San Francisco refuse be composted and used to raise the land level in the Sacramento-San Joaquin Delta region. The delta land, below river level, has been steadily sinking and the compost could solve the problem.

Yakima, Washington: Beef herds around Yakima consume some 538 tons of hay grown on land fertilized only with applications of sludge and effluent from the city's sewage plant digesters.

University Park, Pennsylvania: The experiment begun in 1968 has now become a landmark development for using sewage sludge and effluent to revegetate strip-mined land as well as fertilize crops for town and city officials throughout the world. William Sopper of Penn State University continues to report outstanding results as the project continues. The Muskegon, Mich., system to put effluent back on the land likewise proves a model for other cities to emulate.

Madison, Wisconsin: With increasing speed, grinding is being recommended and used at municipal landfill sites for garbage. Under the leadership of Ed Duszynski, Madison has successfully pulverized refuse—a first step in any composting program—and the pre-treatment is now showing upon more and more sites.

New Brunswick, New Jersey: New Jersey has the largest population per acre of any state in the union. Prof. Charles Reed at Rutgers, New Jersey's State University, is especially aware of the significance of his research to incorporate biodegradable wastes in the soil by the plow-furrow-cover (PFC) technique. With so little farmland remaining in the state, PFC offers some hope to get city officials, plagued with the need for disposal sites, to work with farmers in an effective waste recycling program.

Those are just a few examples of the progress throughout the country. Much research is concerned with overcoming the potentially hazardous buildup of heavy metals which could result in soils after heavy waste applications. Much thought is being given to how to keep those toxic materials from entering the waste stream in the first place. The big advantage of land application is that we can't ignore those toxic substances for as long as we've been ignoring those which were thought to float away and were able to be forgotten.

This could very well be the year the land won back its wastes . . . the year the United States decided to make wastes a national resource, thus contributing positively rather than negatively to our Gross National Product. More companies are entering the field; more legislation is beginning to put the bite on bad practices; more consultants in sanitary engineering are becoming comfortable with the land application concept; more professionals in agriculture and public health are beginning to see the connection between farming and wastes; and, most important of all, more Americans are pressuring for waste management which genuinely relates to environmental limits of resource and energy. The fact is that the first time in U.S. history we're creating a land-waste relationship.

I think we should not try to belabor the point that the marketplace will—through our good wishes, our prayers, or our threats or seminars—produce an endless array of products *because* of the speeches and writings. I do think we should cling, with whatever optimism possible, to the idea that the same economic forces that brought us environmentally-bad products will be the ones to get them out of the marketplace. And that means, one way or another, it will be more profitable to produce those environmentally-sound products.

The organic force that is surfacing in so many different areas of our society has the elements to revolutionize the marketplace for the benefit of us all. How great an impact it will make is the big unanswered question of the moment.

The American Farmer: Folk Hero of the 21st Century

Gene Logsdon

Chapter 2

I like to think I've got the best job in the country. I have an office in the center of a large city (Philadelphia), I live on the outer fringe of the suburbs, I spend most of my working time travelling the rural areas of America, reporting the news of our largest industry: agriculture. Every month I mingle with a good cross-section of citizens. One week I may be in New York, talking to a book publisher; the next week in a place like Thief River Falls, Minn., listening to farmers and small town businessmen complain about the price of wheat. Come Saturday, I'm home in my garden, commiserating with neighbors about crab grass and high taxes. I see a lot of what's happening.

One of the things I see is a glaring communication gap between food producer and food consumer. Understanding between the two groups seems to decrease in direct proportion to the number of words printed in magazines and newspapers to "explain the situation." The urban blue collar laborer often considers the farmer a fellow worker and is shocked to hear farmers criticize labor unions. When city students stage an Earth Day rally wearing gas masks to protest air pollution, a Montana rancher whose whole world knows only clean air, thinks the kids have lost their minds.

But oddly enough, whereas in years past it was the farmer who was considered the hayseed or bumpkin because of his ignorance of city ways, today the farmer, blessed with good highways and television (if you call that blessed!), knows much more about cities than the urbanite knows about agriculture.

Sometimes when we editors at *Farm Journal* can't think of a better way to fill a page in our magazine, we amuse ourselves by interviewing New York cab drivers, Chicago Playboy bunnies, San Francisco Jesus Freaks, or Philadelphia computer wizards about farming. We'll run the survey under a title like "What City Folks Think of You." Always good for a few yuks. Once I actually found a fellow willing to make a guess when I asked him what he thought a self-propelled marshmallow picker cost.

What we find generally with such a survey is that the typical urbanite thinks food is something you eat as little of as possible to stay slim—unless you are flush and can afford a good restaurant meal. Where food comes from is not only a mystery, but a mystery they aren't interested in hearing explained. As a St. Louis plumber put it: "Why should I care about farm problems? The farmer doesn't give a hoot about mine. I'll take care of myself; he can take care of himself."

That view is pretty typical, but I think it's changing. And I hope I can convince you that it should change because it's not in *your* best interests as a city dweller. To explain why, I'm deathly afraid that I will sound like a college lecturer in economics and history, God forbid. What I want to do is present you with a big IF. IF you want to enjoy an America where there are still pleasant places to live, where you can still afford to buy high quality food, where the view out your window isn't always someone else's window, then there are certain matters you need to mull over. I don't know exactly at what age one comes to the realization, but eventually, each of us sees that life is very short and accomplishments rather ephemeral. If a person can't eat and sleep well on the journey, then what the hell is the use of it all?

The first thing you need to understand, as far-fetched as it sounds, is that the farmer is not the master of his own destiny, even though he would be the last one to admit it. Most farmers continue to be farmers because they believe they are their own bosses. Actually, farmers make up only a small minority of the population—less than 3 million and even that is stretching it a bit. There were 3,821,000 farms in 1961 according to U.S.D.A.; down to 2,831,000 in 1971. Agriculture has lost most of its political punch, or at least is in the process of losing it. Many people predict that we are not too far away from the time the United States Department of Agriculture will be only a memory.

21

(And some people, including myself, think it would be good riddance. The department's most important activity now seems to be to swing rural votes toward whatever administration happens to be in power at the moment. This is not the opinion of the magazine I work for.)

But there's a more important reason why farmers don't really call the tune to which they must dance. Historians are finally getting around to the fact that agriculture never has been either the beginning point of America's social order, nor an independent business sufficient unto itself. The settlement of America has always been by economic complexes of town and country, happening more or less simultaneously. You cannot have a commercial farm until you have a market for food, and you can't have a market for food until you have people who do something else for a living besides raising food. The farm depends on the city for existence.

Moreover, there is plenty of evidence to show that the impetus for change in agriculture and even the changes themselves have historically come from cities, not from agriculture, odd as that may seem. (Jane Jacobs develops this theme in her excellent book, *The Economy of Cities,* portions of which are reprinted in this book under the title, "The Importance of Cities to Rural Living.")

You, the urban dweller, have a direct stake in what happens to agriculture in the next fifty years. And you have the power to make it happen the way that will benefit you the most—which, I think, will also benefit agriculture the most.

What do I mean by "benefit?" Let's talk about New Jersey. We could talk about Long Island, or all of eastern New York, or most of New England, or the San Joaquin Valley of California, or Los Angeles County, or northeastern Ohio, or the Milwaukee-Chicago metro area, or the Boston-Washington megalopolis, or any other place where cities are stretching out into the countryside. What all these areas have in common, among other things, is a rapidly diminishing number of farms.

But let's use New Jersey for an example, because there the leaders are willing to go on record as saying that with the way things are going, New Jersey could lose *all* its farms by the middle of the next century.

That would mean the loss of billions of dollars to New Jersey.

That would mean the end of most open space in the state, except publicly-owned park land.

That would mean that New Jersey would, in essence, become an importer of its food.

That would mean that New Jersey residents would pay a higher price for food because of added processing, storage and transportation costs.

That would mean that New Jersey residents, suffocating in a place like Camden or Jersey City, could no longer go for a ride in the country because there wouldn't be any country to go riding in. Nor could people do what has become a traditional popular activity in New Jersey: drive out to genuine roadside stands and buy truly fresh fruits and vegetables.

That would mean that New Jersey would become an even greater Center of Industry (as their Chamber of Commerce seems to desire)—a great smoking monument to the GNP. The whole state would become one vast extension of Newark, now commonly referred to as the "tailbone of creation."

The only spot left where a man could go to find clean air, or the beauties of nature, or himself, in New Jersey, might be the Pine Barrens. And even this last stand of pure air and water in the state might disappear: it's being proposed as a likely site for an international jet-port.

Think about it.

If you ask a city dweller where he'd most like to live, the vast majority will opt for "a place a little out in the country with a few neighbors, some pleasant open space, and not too far from work." They'd like room for a garden, some woodland to walk in, maybe a stream flowing by the place, room for the kids to run and play safely. We know this is true because city families, when given the chance, have migrated to the suburbs in droves in the past 25 years.

But here's what has happened. A developer buys a tract of farmland overlooking more farmland and builds a few houses on it. He knows what people want. He calls his development, "Green Acres," or "Deer Run," or "Old Orchard," or some other bucolic-sounding name. The houses sell fast to people hungering for the environment of the country. Encouraged, the developer builds more houses on surrounding tracts of land (the lots getting progressively smaller as he tries to make a good profit on land he now has to pay more money for) and more people come to buy them. Because of the way our outmoded property tax laws operate, suburban development drives the price and therefore the tax rate of surrounding farm land to levels much higher than a farmer can afford. His land is valued not at $500 an acre, what it might be worth as farm land, but at $5000 an acre, what it might be worth as subdivision lots. Nor can the farmer buy more land at the going rate for farming, to enlarge his holdings so that his gross income will cover the extra tax cost.

The new neighbors from the city, who pined for country life, complain of manure odors that might emanate from his barn at times. Children of the newcomers, untrained to country ways, harrass the farmer's cattle, break down his fences, steal his apples, strawberries and tools.

Eventually, the farmer is forced to sell out. Finally the whole area becomes a facsimile of just what the urbanite attempted to get away from. He loses what he desires in the very act of attainment.

I get hundreds of letters like this one from a farm family in Florida: "In Dade County, 25 miles south of Miami, there are many pick-your-own fields of sweet corn, pole beans, peppers and tomatoes, attracting the city dwellers and the tourists off the four-lane highways to pick fresh produce daily. But the land here is being devoured by shopping centers, expressways, and ecology preservation. Future farming here is seriously jeopardized. What does a grower do, who has three sons wanting to continue in farming? Do the cities want to drive us all away?"

24

Driving farms out is what's happening everywhere as the tentacles of cities reach out to close hands with the tentacles of other cities. And who will be the biggest loser?

You, the urban dweller, will be, and I don't know a better way of explaining that than quoting from letters I received from farmers, after we ran, in *Farm Journal,* the remarks of an urbanite who told us that "all farmers do is bitch, while the government gives them money to stay in business."

"Dear sir," wrote back an angry farm wife in retaliation. "Let me tell you a thing or two. I just returned from taking my husband farmer to the field. He's been up and working since 5:00 A.M. and he won't be home until 10:00 P.M. It's hard work earning what you call those federal farm subsidy welfare payments and enough besides to make a living. People like you make me retch, sir. Talk about bitch. If you put your money where your mouth is—you and others like you—we wouldn't need those "welfare payments". We'd get an honest price for our products. But no. You bitch about the over-the-counter price of food and you don't think a thing about the man somewhere working his guts out so that you can be one of the world's best fed people. Do you know you pay only about 18% of your earnings for food compared to around 50% in other parts of the world? (She's right.) Do you know that the prices farmers receive for their products has not risen appreciably in 25 years, and in fact has often been lower than 25 years ago, while the farmer pays for his inputs triple or more what he paid then?

"We don't like to work a 12 to 14 hour day to make a living. But we do it because we want to stay in farming. But we can't last forever. You better hope people like us *do* stay in farming, because if we are forced out, and corporate farms take over, then that would mean unionized labor in agriculture. When the corporate farms would begin paying union wages, then, sir, you'd really bitch about the price of food.

"Farmers aren't the only people hurting. Small town business is feeling the squeeze too. Move over, sir, a great many more farm and small town "dropouts" are heading for your city. They are good workers and will take over some of *your* jobs. And you have enough worries without adding more unemployment to the list."

Another letter, same subject:

"If Mr. — thinks the bitching is all done by farmers, he ought to check his newspaper. Every day, another company goes on strike and prices go up. If farmers started bitching like this, your stomach would soon be growling. And if prices were raised in tune with things like cars and colored TV sets, your pocketbook would be growling too."

And another:

"If Mr. — would do a little research on Federal payments, he would discover that the "city" farmer, like some of our millionaire Senators, draw most of those funds. As far as the U.S. Department of Agriculture is concerned, the whole department could be done away with, and the regular family farmer would never miss it. That's how much help it is to us.

"Senator Talmadge in a speech recently said that if farmers continue to decline like they have in the past, there will be none of them left 20 years from now. If, sir, you can last for 20 more years, you have made it. After that, no more farmer bitching, not ever."

And another: "Dear sir, would you work at 1950 wages and pay 1971 prices for cars, clothes, and doctors. We do, but we aren't supposed to complain about it. Yes, maybe, 60% of us had better either go on welfare and let the working man support us, or go find jobs in the city and watch unemployment rise a little higher. *I suggest that you and your friends keep looking over your shoulders as more of us throw in the towel and move to cities because we may be coming to take your job.*"

Just this morning (February 22, 1972), I talked on the phone with an executive of one of the biggest and fastest moving farm machinery companies in America. "Looking ahead, ten to twenty years," I asked, "Are we really going to see a continued decline in the number of farmers?" "As far as our sales are concerned," he replied, "the answer is yes. We see fewer farmers, fewer farms, but bigger ones. We're gearing up to serve the big boys who will be left."

26

The big boys who will be left. Already about 800,000 commercial farmers produce some 70% of the food in this country. Will half a million produce 90% of it twenty years from now?

Economists classify farms into five different groups according to gross income: Class I contains those farms grossing over $40,000; Class II farms gross between $20,000 and $40,000. Class III, $10,000 to $20,000; Class IV, $2500 to $10,000; and Class V, under $2500.

Historically of course, there have been more farmers in the three lower classes, but since about the middle 50s, the bigger Class I and II farms have been increasing—a fact which some economists use to show that agriculture is not a depressed business at all, but one showing good economic growth. The decrease has come in the so-called lower classes of farms.

So, in addition to the decrease in farm land from urban development, we've had another and more significant decrease in small farms due to economic pressure. Here's how it works.

Let's take a specific 640-acre section in central Illinois—good farm land, not marginal land. Pretend it's 1950. There are three farmers, (A, B and C, we'll call them) farming that section, each with about 200 acres of diversified crops and livestock. (In 1930, there were probably 8 farmers on that section, each with 80 acres.) The three farmers, A, B and C, are each producing a quantity of food for market at a certain cost, and making a decent living. Then, through the 50s, costs begin to rise while prices remain the same or go down, (which they did).

Despite all kinds of so-called government help, this situation continued. The three farmers could not raise their prices, they could cut costs only so much (making them the most efficient businessmen in this country, by the way), they could raise production per acre only so much with their traditional crops, which higher production only brought on surpluses of food and lower prices. When government tried to hold prices up artificially, it succeeded only in encouraging more production, causing still lower market prices. When A, B and C had reached the limits of their

27

abilities and could still not make enough money, they had only one other alternative: to expand. Since all farmers were facing the same problems, it was mathematically impossible for all of them to expand, even if all of them could afford to. Someone had to go out of business. In the case of our example, it was C who had borrowed heavily to get into farming in the first place, and who did not quite have the same management ability as A and B, who felt the crunch hardest.

So, C sells out. A and B bid on his farm (sending the price rocketing upward) each realizing he needed it. A makes a bigger offer than B can afford to make and gets C's farm. He goes in debt to do it, which means that C's farm must, in effect, continue to make the interest for A's lender plus a profit above other costs for A—*just as it had been unable to do for C.* But A now spreads his per acre cost over a larger area which lowered unit cost, and by socking more fertilizer, chemicals and bigger machinery into the fray, he stays in business.

But costs keep spiralling while prices remain, on the farm, at the same or nearly same levels on through the 60's. B is eventually forced to sell out or change to a different kind of farming—a high value per acre crop like strawberries in which he is quite likely to fail, since the typical corn farmer knows no more about commercial fruit and vegetable marketing than he does about selling golf clubs.

A finally buys not only B's land but the farmer A's land in the *next* section, which makes our hero, A, a Class I, 1200 acre farmer so heavily in debt that he couldn't get out of farming if he wanted to. He has a hell of a good gross income and a hell of an interest payment, and he is, as we say, under the gun. He must use every labor-saving machine and chemical he can get his hands on, or he will go under—and in some cases more than likely take his local bank along for the slide. But all is not lost. This fellow's "worth" has risen dramatically because of the rise in the value of his land. In 1950 his land was worth $200 an acre; in 1970 his 1280 acres runs something like $600 per acre or more—that's $768,000 smackerolas, friend, which ain't bad. But remember, our hero A hasn't got it paid for. If he sold out today, he might not have enough left over free and clear to keep his family in pork and beans for two years.

Farmer A—the half-million or so Farmer A's in the country—are the ones who bear the onus of ecological recrimination, today. Or they should be the ones, because it is they who are producing the bulk of the food. Whether rightly or wrongly, the A's are being accused of mining the soil, polluting the rivers, and chemicalizing your food. I do not intend to defend him—here—but I do want to explain him. He is doing only what the pressures of economics and population have forced him to do. In a very literal way, his accusers are biting the hand that feeds them. To demand that Farmer A change his way of operating is to demand a new social and economic order and these things don't happen just because one demands them to happen. And to try to regulate the commercial farmer by government action, will unfortunately, only run the few remaining small farmers out and increase the number of very large farms.

There is, however, one fallacy in the urban mind (indeed one shared by many farmers too) which needs clarification. There's a great hue and cry right now over corporations taking over agriculture. The truth is that *so far,* less than 1% of the food produced in this country comes from big corporate farms. As a matter of fact, big corporations are going out of agriculture right now faster than they are going in. Reason? First of all, there's not enough money in farming. The profit picture isn't good enough to support the inefficiencies and high wages of the ordinary corporate labor structure. (Which between the lines says something louder than shouting words from rooftops.)

The second reason that big business cannot quite make a go of farming so far, is that agriculture is at the mercy of uncontrollable and unpredictable natural conditions—like the weather. Producing corn is not quite like producing automobiles.

Moreover (reason number 3), industrialists have found, to their surprise, that farming, even under the most modern conditions, demands a certain touch of craftsmanship—the skilled love of personal attention—that you can rarely get from hired labor. A conglomerate sending its cadres of experts into the farming field is a bit like asking a foreman and assembly-line crew to produce a batch of poetry. Won't work. This is agriculture's main weakness—and its ultimate glory. It demands Art as well as Science.

Will the kind of cannibalism in agriculture which has so decreased the number of farmers continue to some kind of logical absurdity where perhaps only two farmers will remain—one owning all the farmland west of the Mississippi and the other all the land east of it? One thing seems certain. As the land becomes the possession of fewer and fewer people, democracy as we have known it, as Jefferson foresaw it, will cease to exist. I can find no example in the history of humanity where a democracy, or even a strong republic, lasted long when the land was owned by an oligarchy of the very rich.

However I do not believe that the trend toward fewer and bigger farms can be stopped by any of the conventional ways being tried. Nor do I think that they necessarily *should* be stopped by these conventional methods, because the cure would be worse. Besides you, the consumer sorely needs a healthy number of large farms. But the point I want to make is that the ways implemented so far in an attempt to halt the attrition of farmers out of agriculture, attempts often heralded piously as "saving the family farm," are not only ineffectual but dangerous.

What am I talking about? Government regulation and general farm organization. Neither has been able to "save the family farm" in 50 years of concerted effort and billions of wasted dollars. Government attempts to do so have meant spending vast sums of your tax money to the gain of only the larger farmers. Farm organizations, seeking a higher profit for their member-farmers, have succeeded in doing just about everything except that. Right now, farm organizations are pushing hard for laws that would give them the same kind of monopolistic bargaining power as the labor unions enjoy. I can find, in my travels, few farmers who favor these moves, and the moves certainly must be looked upon by you, consumers, with much suspicion.

But rather than air my own opinions, I will let a real farmer talk on these points because what he says reflects what most good small farmers believe, and because he gives you, the urban dweller, a view of agriculture you rarely hear about. His name is George Mueller, and I take his words directly from the Congressional Record as he testified in Washington *against* farm bargaining legislation being proposed by farm organizations.

"It is a misconception to think that we need to bargain for higher prices if we are to save the family farm. This is the old claim that we also heard to justify higher price supports. A look at the record indicates that the small family farm actually disappeared at an alarming rate during our period of guaranteed price supports. Why? Because if you guarantee a stable price you are giving to the large farmer the stability he needs before he can invest heavily in labor, equipment, land, and buildings. The small farmer has a better ability to ride out periods of low prices. Stable and higher prices assured by bargaining negotiations and quotas will give just one more advantage to the large efficient farmer over the small producer. We have seen this happen with government price supports. This is why it is large growers who are leading this fight for increased bargaining power. They each have a large investment to protect. . . . Bargaining will not help the small grower. In fact it will put him at a greater disadvantage. . . "

"The point we want to make is that *farming has been good to us.* The opportunity to compete under a free enterprise system in agriculture was open to me as a city boy. No one said to me that because my father had not been a farmer that we could not farm. No one said because there was too much milk—and there was too much in 1960— that we could not produce milk.

"When we entered farming we found the industry open to all and very competitive. For this we are grateful. This is why we are here today, my wife, kids, and myself, we want to keep agriculture open and free . . . under a competitive agriculture, we have enjoyed the freedom to follow the desires of the market. . . . It is the freedom of price to move up and down in our markets that has directed progress in our agriculture. . . . We all have seen our merchant marine, our steel industry, our electronics industry, our textile industry and many others price themselves out of world markets and their own jobs, due to bargaining demands of organized labor. . . . Up to now we on the farms have concentrated on productivity per man. As a result our agriculture is alive, competitive, flexible, and free. Both the consumer and the farmer have benefitted."

The urban dweller needs to be aware that there are many smaller farmers like George Mueller who are doing quite well in agriculture, thank you, and that they are ready, at any time, to answer any needs that you, the consumer, place in the marketplace. If you desire any certain type of food, quality of food, variety of food, you have but to make this desire known with the vote you have in your pocketbook, and the small American farmer—perhaps one city-born like George Mueller, will not only stay in business supplying that need, but he will begin to proliferate, not decrease. It is the urban consumer who has it in his power to make or break any particular kind of farm operation.

Near Norristown, Pa., itself a suburb of Philadelphia, there is a small dairy farm completely hemmed in by subdivisions. This dairy's success in remaining viable in an urban, residential district provides the best example I know of how farmer and consumer can exist side by side to the benefit of both.

We'll call them the Brown Brothers, the owners of the dairy. They decided they did not want to move out of the community where they had lived all their lives. They felt that by selling off some of their land and concentrating on producing milk only, they could keep farming on expensive, high tax land, *if* they sold their milk retail rather than wholesale.

A simple decision, you would think, but the brothers immediately became involved in a struggle with the vested interests of the milk industry in Pennsylvania. The big milk companies did not want upstart farmers to start retailing milk because *it was a well-known fact that a farm retail operation could make a good profit at a lower price than those set by the Milk Control Commission.*

32

The battle that followed was too long and complex to recount here. But eventually, the brothers Brown won the right to produce, process and sell their own milk at their own price.

And how the suburban neighbors responded. The "Brown" brothers have never had any trouble selling their milk; their problem has been to produce enough to meet the demand—the dream all farmers and farm organizations have tried unsuccessfully to accomplish with rules and "cooperation". People flock to the Brown operation to watch the cows being milked; literally thousands of school children visit the farm annually to see the animals and to learn something about where good food comes from. Neighbors love the beautiful green, rolling 15 acre pasture full of cows next to them. They take a motherly interest in the herd. If a cow is calving in the field, the Browns will get phone calls from concerned observers. No one complains of manure odors when the Browns spread the old barn bedding on the fields: the odor is noticeable only a couple of days a year, and people know there is no better way to keep "their" pasture green and fertile. Many of the Browns' customers buy the manure by the bushel basket for their gardens—another source of income for the Browns, which is mutually beneficial for the whole surrounding environment.

Neighborhood children are trained like farm children—to be responsible for private property. They do not break down fences; they do not chase the cows.

But best of all, the people can buy good milk at a cheaper price than they would pay in the supermarket.

There is no reason why operations like the Browns' cannot increase and multiply, just as the suburbs increase and multiply. Already, many such farms do exist. Some sell eggs. Some sell their own beef and pork. Many, indeed the classic examples, sell fresh fruits and vegetables. *I do not know a one of these farm retail markets, operated by good, honest farmers, that is not making an excellent income. The reason that farm organizations, government advisory agencies, and politicians are not trying very hard to tell farmers or consumers about how to take advantage of this direct food exchange is because its success would make the organizations and agencies obsolete and unnecessary.*

Urbanites who want good food markets close by (for fresh food, the market must be close by) and who want to preserve open space in their neighborhoods in a *practical* way, should know and support the kind of laws that make it possible.

1. Basically, the Brown Brothers can go on operating because of a wise Pennsylvania law which allows them to sell *the milk they produce on their own farm* at their own price, despite whatever monopolistic devices control the rest of the market, so long as their sanitary conditions meet the requirements of local health codes.

2. A second law which enables small retail farms to stay in business often, is a zoning ordinance, which fortunately, is quite commonly followed in many states. The ordinance states that a farm existing in an area zoned residential can sell from that farm any produce raised on that farm, but only produce raised on that farm. Many such roadside stands exist in New Jersey that I am familiar with. They are the very best places to buy fresh produce because you know that the food was raised right there, not shipped in.

3. The property tax, which is mainly school tax, is probably the single biggest small farm destroyer in the country —in densely-populated areas. The constitutionality of the property tax is now being challenged successfully, and if substantial changes come to lift the burden of supporting schools solely on the basis of property ownership, it is bound to mean more small farms in suburban areas, other things being equal.

Some states are already experimenting with tax assessment based on land use, not land potential. That is, as long as land is being operated as a farm, it is taxed as farm land, not as subdivision land, even if you could sell it for subdivision land. This seems like a simple and practical and even logical kind of practice to put into effect. Unfortunately it may not be, due to the colossal greed of men.

4. If there are retail farms in your area, give them your business. You can't go wrong. Prices will be better usu-

34

ally, quality almost always better. The weakness of the small retail farm is that it cannot easily maintain supply in tune with demand. Even a retail dairy farm, which produces milk every day, has a rough time keeping enough on hand. But remember, if you go there and he is out of milk, don't be too upset. The fact that he is out, is proof he is not buying someone else's milk on the side, the source and quality of which you cannot check the way you can check his.

An eastern strawberry grower will have his own strawberries only in season, but they will be fresh, even perhaps organically-grown. At any rate, they will taste a whole lot better than out-of-season strawberries in the supermarket. It's best anyway to eat "in season" and can or freeze for winter.

Tell your small farm marketer that you would buy frozen or other processed fruit and vegetables from him, if he had them. If enough consumers asked, you can be sure that small farm retailers would get into small farm processing, and that in a nutshell would solve almost all our so-called food problems today.

You, the consumer, can save the American farmer as we know him, from becoming the folk hero of yesterday. And you just might save your own environment of tomorrow in the process.

How to Love the Land and Live With Your Love

Ruth C. Adams

Chapter 3

Going South on Route 100, you turn East at Bucktown on Route 23. Driving through five miles of rolling countryside, fine old stone houses and woodlands, you notice bake ovens, smoke houses and tall whirling windmills. Windmills? You cross the boundary into Lancaster County, Pennsylvania, and you are in the eighteenth century.

A vast panorama of magnificent farmland laced with winding lanes opens on both sides of the road. Sturdy, timeless houses and great stone barns nestle in cozy groups, with chicken houses, stables, sheds, poles sprouting huge, old dinner bells. And windmills. The fields are almost unbelievably verdant and fertile. Sculptured in greens, browns and reds, they lie on the lavish landscape wantonly, promising rich, rich harvests in the season ahead.

There is something peculiar about these barnyards. There are no cars, no trucks, no television antennae, no power-driven machinery of any kind. There are no electric power lines marching down these lanes. Are these people too poor for such amenities? No. These are the old Amish. These are the Plain People who live, as they wish to live, departing hardly at all from the way their ancestors lived in the times of the Reformation. The House Amish, the Plain People use no electricity. They own no cars or trucks.

By the time you get to Morgantown you know how they get along without them. You meet shiny, black buggies and wagons pulled by fine, well-tended horses. The Plain People are riding in open or closed buggies or small wagons. If it's Sunday they are visiting friends, relatives or neighbors, or possibly coming from church, which is held in private homes rather than a church building. On weekdays the roads are almost empty, for the Amish are at work in their fields, barns and households. Most of them raise all of their own food. They butcher their own hogs and cattle, milk their own cows, pick their own apples, harvest their own grain and take it to the mill to be ground. Their luxuriant vegetable gardens boast rows and rows of corn, peas, lettuce, beets, carrots, beans, kohlrabi, brussels sprouts, spinach. An Amish housewife may have 22 vegetables in her garden.

Inside the house, furniture is staunch, well-made, un-adorned. The kitchen is the center of family life, the only room heated in winter. An old-fashioned black wood and coal stove provides for heating and cooking. The spinning windmill pumps water into a cistern from which the Amish housewife hand-pumps it into her sink. All the Amish clothing is made by the mother and daughters of the family. It is designed as Amish clothing has been designed since the Reformation. Little girls and their mothers wear long dresses with full skirts and aprons. The bright colored cloth is always plain, never figured. Aprons for everyday use are black. For Sunday and special occasions they are white. The men wear black. Everyone has a Sunday-best outfit. The rest is strictly for everyday. Women wear prayer caps at all times, their long hair in tight, neat knobs at the back. The men have hair to their shoulders and, as soon as they marry, grow beards. Their broad-brimmed black hats (straw for summer) are handsome and lend a dashing air. Boys of four wear the same kind of hat with great dignity. Little girls and women may wear bonnets. Clothes are practical, comfortable, made of easy-to-launder cotton.

The refrigerator in the kitchen may be run by bottled gas in some of the more progressive homes. An old-time refrigerator might be a handsome chest-like affair in the corner, with pipes bringing fresh water from the stream, which pours down through the double walls of the refrigerator, cooling it to 50 degrees or so even throughout the summer. The water runs off into the barnyard for the cattle to drink. The pump at the stream may be an elaborate but simple-appearing gadget which looks as if it runs by perpetual motion. There is no electricity or other power source. Elaborately-balanced water wheels keep the pump working away day and night, so long as the stream runs.

Nothing is wasted in the Amish household. Scraps from sewing bees are worked into quilts for warmth and decoration. Table leavings and waste from the household canning are fed to pigs and chickens. An Amish housewife may can or preserve as many as 1000 jars of fruit, vegetables, pickles, meat. She smokes bacon and ham in the smokehouse. She makes sauerkraut in a big tub. She makes apple butter in a huge kettle.

Clothing is worn until it wears out, then replaced with another piece exactly like the one which wore out. Fashion is never involved. Baths are simple affairs in metal tubs or crockery basins. Nowadays a flush toilet and a bathtub are permitted in some districts, if they are located downstairs so that no electric power is needed to carry the water upstairs.

The Amish are conscientious objectors and participate in no wars. Their hatred of the military is so great that they shun buttons on their clothing because ancient military uniforms displayed rows upon rows of buttons.

Today the Amish in some parts of the country are in trouble with the law. They are convinced their children should not go to school beyond the eighth grade, for they say children do not need any education except for the education that comes from living with The Land. They will be farmers always. They have no other occupation. Higher education is not only unneeded, they say, it tends to corrupt the farmer, perhaps turn him from The Land, which is all-important.

The Amish have an almost obsessive ferocity about The Land. Every waking hour, every thought is given over to the soil and the good things that come from it. They quote Bible references to explain this devotion. Some people find their outlook narrow, their conservative, highly moral religion stifling. And some young folks do leave the Amish community and become "worldly". The number is very small—perhaps two percent.

The Amish have always lived with The Land, in perfect harmony, returning to it everything they take from it, cherishing it as the source of all life. Living with The Land is hard work. You get up before dawn; you work till dark, then you light a candle or an oil lamp to finish the chores. You have the immense satisfaction of seeing the im-

38

mediate results of your labor in the harvest, the cold cellar full, the can cupboard full, the corn-crib full, the hay-mow full—all by the first frost.

Late fall and winter are for socializing in those hours when you cannot spread manure, mend harness or prune trees. Social activities are determined by the distance your horse can go in an hour's travel or a day's. Although most Amish read newspapers and magazines, they like best their own newspaper, *The Budget,* a daily, printed in Sugarcreek, Ohio which is a chronicle of the doings of the 50,000 Amish all over the country. The December 16, 1971 issue carried news like this: "Dayton, Virginia: It doesn't seem like winter as the temperature was in the 60's." "Meyersdale, Pa.: Grass is nice and green and we had our cattle out in pasture most of last week." "Spartansburg, Pa.: Mennos had supper at Dave D. Byler's Thursday evening. They took a load of their belongings in to Geauga Friday and will take another load Tuesday and the rest Wed. They will live in Dan J. P. Miller's small house until they find something else. We are planning to have a singing for them tonight."

News like this is all the 14 pages of *The Budget* contain, except for a few ads for the very few products an Amish man might buy once in a lifetime—a clock, a blanket, a few yards of cloth. One who customarily listens to the six o'clock news of mayhem, disaster, murder, rape, addiction, war, graft, robbery, laced with idiotic commercials for trivia, can be excused for wondering if it might not make for a better life to have no daily news but *The Budget.* These simple family notes are somehow almost all that are required by people who have renounced all wars, all violence, all commercialism, and, it now becomes evident, in an environmentally concerned age—all pollution.

Except for a bit of cloth, enough fuel to heat one room, some few tools and enough hardware to rebuild their wagons and plows, the Amish need almost nothing from outside their communities. So far as their way of life is concerned, we could close down most of our major industries with all their pollution and scandalous waste of resources, and go back to The Land to live. Nor is there any reason to be poor. The Amish have done such an excellent job of living with The Land that, for a considerable number of years out of the last 100 years, they have been listed in Department of Agriculture statistics as having the highest farm income in the country.

They obey all federal regulations about restricting crops, but they refuse to accept subsidies. They refuse social security, too, and have no need for it, as their old folks are well cared for and highly respected as the wisest people in the community. And they are, of course, since they know more about farming than younger people, and knowledge of farming is the only knowledge demanded of an Amishman. No one feels any need for insurance as neighbors will help rebuild a barn which burns, relatives and friends will provide for a widow or orphan, without question or remuneration.

The struggle of an Amishman named Yoder against the might of the State of Wisconsin which has decided he must send his children to high school or go to jail has now reached the Supreme Court. In the January 15, 1972 *Saturday Review* you can read the full story. Says Stephen Arons, "Perhaps the greatest irony of the case is the notion of freeing Amish children from their community. The Amish community experiences little delinquency, causes and fights no wars, uses no polluting machines, eschews materialism and has no economically-based class system. To save these people from the quiet sanity of their lives by forcing them into the center of the psychologically unhealthy atmosphere of modern America strains the definition of freedom beyond recognition."

And the prosecutor, John William Calhoun, speaking for the State of Wisconsin, says, in part, "The issue is not the Amish life-style or a matter of pluralistic vs. egalitarian societies. In fact, to many of us caught in the remorseless, day-to-day crunch of daily living, the Amish life has great appeal. However, as Stuart Chase has said, 'Retreat to a simpler era may have had some merit 200 years ago when Rousseau was extolling the virtues of the Cro-Magnon man, but too much water has gone through the turbines.' "

Has it? Are we going to take the word of one economist and one lawyer that it has? You can live with The Land as the Amish do, if you care enough to do it. You can buy a windmill and an oil lamp in Lancaster County. You can get a horse and buggy, a plow and a cultivator. It is, apparently, perfectly possible to live a successful, fulfilling life with The Land, without any of the blessings of our great commercial-industrial establishment.

The Amish have been demonstrating to us for hundreds of years that all you have to do is care enough about The Land, to cherish it, to regard it as the single most important thing in life, the thing for which you are willing to give all the hard work and devotion necessary to make your relationship with it successful and permanent.

The Land Must Live

Wayne H. Davis

Chapter 4   I was born in 1930 and thus had the wonderful opportunities of a child growing up during the Great Depression. Ours was the magnificent world of work, fun, games, good times and happy children. I never saw a whining, fussy child until the Great Economic Boom brought us Future Shock along with television, swimming pools, built in obsolescence, and finally the use-it-once-and-throw-it-away economy.

The Depression gave the nation its lowest fertility rates. Children grew up with a sense of stability of the land and their environment, a sense which contributes strongly to the feeling of security and the development of sound mental health. We saw first hand the fundamental relationship between man and the land, a land that was not being destroyed to build new shopping centers and super highways. We also learned the basic laws of ecology by simply observing the fundamental interrelationships among living things. We knew what type of place the catbird would choose for its nest and where it would forage for food. Without television and swimming pools to distract us, we were able to spend all our spare time in summer playing outdoors.

Perhaps of greatest value to us kids growing up during the Depression was all the land we had. Around my home there were many lots whose titles had reverted to the city in default of taxes. This land grew up in wild cherry and locust trees. It was used by no one but the kids and it belonged to us as completely as if we had owned title to the land. Here we built our tree houses, kept trails cleared, planted apple cores and peach seeds, cut a tree now and then, made slingshots, and had corn roasts. The land was ours to manage as we chose so as to keep it wooded and reasonably wild. It was large and diverse enough that we always had a wood thrush, a towhee and a field sparrow nesting there in the city. We managed the land so as to

42

retain enough of each of the different habitat types needed by these birds.

As I watched the drastic reshaping of our Earth as urban sprawl engulfed the nation following World War II, I have often wondered what the effect would have been on us kids had the bulldozers moved in and converted our woods and fields into a parking lot for a new shopping center. At the very least such a traumatic experience would have alienated us against the Establishment and caused us to question the values and the concept of progress held by the older generation.

Yet today such an emotional shock has come to be a routine part of the growing up of most children. I suspect that it is also a major cause of such problems as the generation gap, rising rates of delinquency, crime, drug addiction, mental health problems and the general rebellion against society and its lack of values.

Consider the changes in our ability to destroy the land that have developed within one generation. There were no bulldozers when I was a boy. Roads were built by men using a pick and shovel. Coal was mined with the same tools. Today we have giant earthmoving machinery that can pick up 200 cubic yards of earth in a single scoop to get to the coal seam beneath it. In building highways we do not hesitate to move mountains and fill in valleys.

I can remember when it was a major job to fell a tree. In cutting timber two men with broad axes would chop into one side of the tree and then use a hand-drawn cross cut saw from the other side until the tree fell. With today's power saws two men could clear a substantial area in the same time. To destroy a forest and replace it with a paved parking lot is not nearly the job today that it was 30 years ago. This is a major reason why we are now paving the land surface of the U.S. at a rate of over a million acres a year. In California alone it is going under at a rate of 375 acres a day.

Ten years ago I bought a home in Lexington, Ky. Behind the house is a stream and small open field. Although I realized that in the nation's 14th fastest growing city the last of such green open spaces would soon be destroyed, I decided that it would be nice for the children while it lasted.

The little stream died last year. Even though it had been barely limping by from one year to the next, the end came as a shock for me and the children. We had known it well and had come to be close friends.

When we arrived in Lexington I was surprised to find that this little stream was alive. Although it had lived happily for several thousand years before I first saw it, the impact of man had hit it pretty hard by then. It arises from a culvert leading from beneath a parking lot of a shopping center, flows about 50 yards, and then enters another culvert which carries it underground so that the beauty of suburban Lexington need not be marred by a free-flowing stream.

In this short section of stream dwelled an abundance of crayfish, snails and leaches. All three repeatedly found their way into my home in glass jars as the children successively went through this stage of their development. The stream was a laboratory for the study of life, much to the delight of the neighborhood children, and the unfortunate consternation of most of the mothers.

Few care about the loss of this tiny stream and others of its kind. But I think its loss is an important symptom of the disease that affects this dying Earth and represents a cause of our ever-growing social degeneration and loss of humanity among mankind. The next generation of children will never see a living thing in this stream. As they grow up with stinking, dead and trash-filled streams, would you expect them to develop an appreciation of life and a humane attitude toward animals and their fellow man? I don't see how they could unless they rebel against the ugly, meaningless and self-destructive life style that my generation has left for them.

There are two other messages that come from my little stream to you. First is that if you want your new baby ever to be able to fish in the streams you had better join the fight for clean water now. Life is disappearing from our watercourses at a rapidly accelerating pace.

Second is a broader message from Earth to all its human inhabitants. Earth is giving us a non-negotiable demand. Either we change our ways or she will roll over and die. If we continue to destroy all life on successive pieces of land and water on this finite Earth, the end result is known. It is only a matter of time until we have destroyed ourselves.

Next to go will be the open field behind my house. I would love to buy it as an investment in the mental health and stability of my children and future generations. I would plant a vegetable garden out there, but most of the land would be undisturbed, to grow grasshoppers and lightning bugs.

But I cannot buy the land, for it would cost about $100,000, which is many times my annual income. The irony is that in the depth of the Depression my father was able to buy a piece of land across from our house to prevent someone's building on it. He paid $150 and $200 per lot, or about one tenth his annual income, for it.

Thus we see that, although modern mass production and a growing economy has given us more material things and has allowed us to turn our natural resources into junk at an ever accelerating rate, we are now less able to afford those things needed to maintain the quality of life than we were during the Depression.

The reason we have been gulled into thinking growth is progress is that we have a basic misunderstanding of the difference between cost and value. When speaking of land and homes, we use these words as interchangeable synonyms whereas they are actually antonyms. This was best brought out by naturalist writer Joseph Wood Krutch in the last paper he wrote before he died.

Twenty years ago Krutch built a home on five acres outside Tucson, Ariz. As the city crept out to surround his little oasis in the desert wilds, real-estate developers tried to get him to sell his land, telling him that it had risen in value to many times what he had paid for it.

"Value for what?" he asked. "Certainly not as a place to live." Now that the traffic congestion, noise, smog, high crime rate and drug addiction had come out to join him, he claimed the value of the land had gone down. The developers found it hard to believe that a man would rather live on a piece of land than to sell it at a profit.

What goes up is not value, but cost—cost to everybody. The owner does not gain by selling, for to get a comparable home he must pay the same high cost plus the agent's fees.

Cost goes up because there are more and more people each year entering the housing market. Since they are not making any more land these days, but are destroying it at an ever-increasing rate, this means that each year more people are snapping at an ever-smaller piece of the pie. Naturally the price goes up.

So what has developed as a result of urban growth in America is an inverse relationship between the cost and value of land and homes. Cost goes up, especially for young people trying to acquire their first home, and the value goes down for everyone, as the quality of life deteriorates in a cloud of smog, noise, traffic and the other unpleasant by-products of growth.

An earlier writer who recognized the value of land as different from the cost was Aldo Leopold. "There is much confusion," he wrote, "between land and country. Land is the place where corn, gullies and mortgages grow. Country is the personality of land, the collective harmony of its soil, life and weather. Country knows no mortgages, no alphabetical agencies, no tobacco road; it is calmly aloof to these petty exigencies of its alleged owners.

"Poor land may be rich country and vice versa. Only economists mistake physical opulence for riches. Country may be rich despite a conspicuous poverty of physical endowment, and its quality may not be apparent at first glance, nor at all times."*

The land behind my home is zoned for agricultural use. Land bought for that use sells for about $300 an acre. With a zoning change for residential housing use, the cost jumps to $15,000 an acre. Changed to professional or busi-

*A Sand County Almanac. Sierra Club/Ballantine. New York, 1970.

ness use, the cost of the land would take another quantum jump.

That the highest value of the land, on the other hand, is as open space, is evident from the unanimous desires of the people of the neighborhood to have it remain as such. The people of this city do not need more office buildings, more service stations and drive-in restaurants, or more traffic problems. It is ironic that, with their demand for housing, new people moving to the cities are paying for the destruction of those values that we most wish to keep. And the cost of our housing is so high that we of the affluent society cannot afford to buy additional land in order to save it.

In 1949 Aldo Leopold wrote that we must develop a land ethic. The truth of his message comes across more urgently with every passing day. The destruction of the land for personal financial profit is a behavior that is simply not ethical and cannot long be tolerated.

Ethics in human behavior have probably been with us since our earliest civilization, although what we consider ethical has changed somewhat with time. When Odysseus returned from the wars of Troy as hero of the wooden-horse episode, he hanged on one rope a dozen of his slave girls whom he thought to have been unfaithful during his absence.

There was no question of right or wrong, for no ethic is involved in the manner in which a man disposes of his property. The act would be considered wrong today only because people are no longer slaves and women are finally being recognized as individual people entitled to the same personal consideration as everyone else. But the lack of relationship between morality and that which the law considers private property remains unchanged since the time of Homer, and is a major factor which prevents any real progress in dealing with some of the most obscene aspects of the environmental crisis.

Today's equivalent of the Greek slave girl is the land. This land is your land and my land, and like air and water it is a basic element upon which our survival depends. On it stand our cities and our homes, and from it we derive nearly all our food. And the land, more than any other factor, is the measure of the quality of life now available to you and me. It is the landscape, and the trees and the rabbits thereon. It is also the city park, or the one that should have been there. For the farmer and his wife it is peace of mind as they gaze out over their fields on a quiet evening.

Yet, as we enter the last part of the 20th century we still have no land ethic. Although we sing our praises to the amber waves of grain, we consider the land upon which it grows as no more than dirt, to be sub-divided and covered with asphalt as soon as the profit margin becomes satisfactory. Land is treated as a commodity, to be exchanged for money, with the sole object being profit for buyer and seller. The fate of the land at the end of the hangman's rope is supposed to be of no concern to anyone in spite of the fact that it is of value to all mankind.

No one on earth is more deserving to stand tall and proud than the American farmer, who has fed you and me and most of the rest of the people on Earth. Yet farmers are kicked around economically and forced to flee to the cities by the thousands each year to earn a living while the giant land companies get an ever-tightening stranglehold on the nation. It is a sad commentary that the poorest investment one can make is to buy a farm and farm it, while his best investment is to buy a farm and subdivide it into a real estate development.

To the ecologist perhaps the most insane of all meanderings in the business world is that which equates land to paper money and makes the two freely interchangeable. Once sold, the land can be flooded behind a dam, blasted away to open a mine or road or sent helter skelter down into the streams via erosion. We consider no moral question to be involved in this. The only value judgment for society is made on the other side of the transaction: The money received for the land is viewed in a positive light as a stimulus to the local economy.

Perhaps, the time is ripe for a new religion based upon a reverence for the land and all the life which springs therefrom. Is it not every bit as logical to worship the solid earth beneath your feet from which the mystery of life springs eternal with each vernal equinox as to have to imagine some unseen being in the sky? If the Great Spirit still stalks this Dark and Bloody Ground so long after the last Indian fled to the West from his Happy Hunting Ground ahead of Boone and Kenton, surely He must reside in the land and trees as He always did. If we made the land our temple, perhaps at last our fellowman will respect it as he does the man-made structures with their spires rising helplessly in the wrong direction in their search for hope and eternal life.

The time has come when we must all develop respect for the air, water, land and the living things thereon. We are all dependent upon the same life support system of Earth and must protect it if we are to survive.

I hold these truths to be self evident: All living things are created equal and are interdependent upon one another. All flesh is grass. Only plants can make food. Man and all other animals are totally dependent upon the plants which we so casually push aside in pursuit of the ever greater megalopolis. Animals need their oxygen and the plants our carbon dioxide. Both are dependent upon numerous species of microbes which make amino acids and vitamins, digest food, fix nitrogen for our use, and return it to the air when we die. And all are dependent upon the exceedingly complex ecosystem of producers and consumers, predators and prey, herbivores and carnivores, and parasites and disease, to provide for their needs for survival and to control their numbers. Man cannot survive alone. Nor can he continue to increase his numbers at the expense of other living things.

But man is arrogant. He refuses to face reality. Four centuries after Copernicus he still really believes that the earth is the center of the universe and that God's only concern is with his welfare. A century after Darwin man still thinks of himself as apart from nature, with a divine destiny to conquer nature and exploit the other creatures for his own use.

No wonder our young people rebel. They are concerned not only about the deterioration of the quality of life, but also about the prospects of surviving on the dying planet they will inherit. Young activist Penfield Jensen recognized the real issues when he wrote, "We will stop the destruction of this planet even at the cost of our own futures, careers, and blood. The situation is simply like that. If you are not going to live for the earth, what are you going to live for?"*

*No Deposit, No Return. Huey D. Johnson, Ed. Addison-Wesley, Reading, Mass., 1970.

New Technological Remedies for the Farm

Robert Rodale

In my opinion, the land grant colleges have helped to foul up this country by applying too many simplistic technological remedies to farm problems without trying to foresee the eventual consequences of those remedies. Workers at the land grant colleges have continually used advancing technology to replace human hands with machines, chemicals, and special varieties of crop plants. The result has been more food produced by each farmer and on each acre, but at the same time much displacement of people to the cities, high costs for welfare, other social disruption, and often sad environmental consequences.

In using the power of advancing technology in such blind ways, the land grant colleges and their allies—the chemical and machinery firms—have not done things differently than other segments of industry. Almost all phases of American life for the past 100 years have been characterized by such technological penetration, with little thought for what is likely to happen beyond this year's profit and loss statement. The automobile industry is a perfect example. All it appears to be concerned about is the production of more cars each year, plus the making of more highways on which those cars can travel. The basic question of how people can transport themselves in the most environmentally sound, economical and satisfying ways appears not to be the concern of the auto industry. That is a problem for someone else to solve, they seem to say.

Food is another example. Technology, blindly applied, has given Americans a fantastic range of convenience food —and nutritional problems that were not dreamed of before the advent of that technology. The same kind of indictment can be—and has been—made of many other facets of American life, and steps are now being taken to try to correct those problems.

I believe, however, that the problem of wrong use of

technological remedies is more serious in agriculture than in industry and other phases of life, and merits special attention. There are several reasons:

1. The government, through the land grant colleges, has been the primary agent for this technological disruption of our lives and environment. Therefore, government has a special reason to try to set things right. Also, because the land-grant colleges are under government control to a large extent, the means for changing the direction of their work exists.

2. Agriculture, rooted in our fertile soils, is the basic source of American strength. Technological mistakes and the disruption of our farm population sets the stage for a serious long-term threat to our nation's health. The technological manipulation of our agriculture is a perfect example of the all-too-human trait of putting short term profits before the obligation to maintain resources for long-term use. Chemical agribusiness is not proven as a long-term technique. It is still experimental.

3. Finally, the vast rural lands of America have traditionally been a refuge for our troubled citizens, seeking new opportunities, and a new start in life. The present system of farming, oriented to big business, has effectively closed off that alternative for millions of people, and will shut it off entirely for all but a handful of farmers if the present trends continue. If that happens, one of our most precious social resources will have been lost, replaced by urban ghettos of the most miserable kind.

My constituency, the organic gardeners and farmers, are the remnants of the many millions of people who at one time constituted the yeoman core of American stability and strength. We are largely the little people still living on the land, not the businessmen farmers. We grow vegetables and fruits on small plots, using natural and non-chemical methods because we have found by experience that those methods are very effective. We concentrate on building the fertility of the soil, because we know that a fertile soil produces abundant crops with much less work and expense than a depleted soil.

There are some farmers in our organic group, and more are joining every day, but in the farm country we are still a tiny minority.

The amount of help that the land grant colleges have given to the organically oriented people over the years is hardly large enough to be worth mentioning. Some of the techniques of modern, conventional gardening and agriculture are used on organic gardens and farms. Improved tractors and tillers are a help, and so are the new biological controls for insects. But the great bulk of new chemicals and machines and ideas coming out of the land grant colleges have been anti-organic in their orientation, and of no use to us.

The real tragedy is that the agriculture colleges have often attacked the organic people—who really are the only farmers and gardeners completely in tune with the environment—simply to create a smokescreen to mask the stupidity of their own technological policies. We are the kooks and the nuts, they say, while their chemical-spraying farmer, sitting on his mammoth tractor, is supposedly nature's nobleman, wisely following their scientific instructions to the letter.

Without really knowing what organic growing techniques are, and with even less knowledge of how to use them, they repeat the bald statement that millions of people would starve if organic farming was universal.

The real truth, which these land-grant college scientists don't want to face, is that if organic systems were used universally in agriculture and in urban life, our country would be much better fed and stronger in many ways. But you cannot just take the chemicals away from conventional farmers and expect them to become effective organic farmers overnight. You must have a plan, and do many things in an organized way.

Garbage, sewage and other organic wastes must be returned to the land instead of being burned or buried. That would solve an important urban problem.

Displaced farm workers now living in cities must be given the chance to return to the land with dignity, working their own small plots of land where they can support

53

themselves. That would save billions of dollars in welfare costs.

Most importantly, the land-grant colleges must use their scientific resources to create a new generation of what I call the soft technology of farming. They must create machines and techniques that are better and smaller at the same time, instead of concentrating on large-scale techniques that always end up replacing people. We organic people do not want to go back to the old ways. We are not advocating a return to primitive farming, where people are worn out by hard work by the time they are 40. We want a new, ecologically-oriented agriculture that can be made possible by the intelligent application of the best scientific thinking to our problems.

Here are some of the areas in which scientific effort is needed:

1. Energy. Conventional farm technology is essentially slanted toward making the farmer an agent in the use of stored solar energy (in the form of processed coal, oil, gas, and soluble fertilizer deposits) for the increased production of crops and animals. By contrast, all farming prior to 100 years ago, and organic farming today, operates primarily on current solar energy falling on crop lands.

Absorption and conversion of current solar energy is far from complete using present methods. Through photosynthesis, plants convert only a small fraction of sun energy into usable food. By extending the growing season through natural means, ways can be found to increase the conversion rate of current solar energy on small farms. More intensive methods for growing fruits and vegetables also make much more efficient use of the sun's energy than does the growing of most farm crops, such as wheat, corn and soy beans.

With new technology based on more scientific input, sun energy can also be used on small farms for home heating, waste conversion, and increased movement of water from the subsoil to the surface, by way of deep-rooting plants. Also very interesting is the culture of semi-tropical fish (eating low-priced grass as food) in solar-heated dome structures.

Other sources of energy can be tapped for small-farm use. Wind-power generation can be perfected, and organic wastes can be used to produce methane gas for heating, lighting, and even for powering of automobiles. Power storage systems suited for small-farm use can also be developed.

2. Waste conversion and fertilizer production. Ways can be developed to make many waste products of urban living into valuable fertilizers, with less labor and handling than is currently needed. Present technology is adequate to convert almost any organic waste to a fertilizer or soil conditioner, but process-costs need to be reduced. Also, subsidies by urban government seeking to dispose of wastes should be directed to small, organic farms.

3. Machinery. Agricultural engineering departments of land grant colleges should cease work (at taxpayer's expense) on machines for large farmers and work only on machines that will make small farming more practical and competitive. The rotary tiller is such a machine. Using small power units, it enables large-scale gardeners to do a thorough job of tilling the soil. It is essentially a miniaturization of the traditional plowharrow machines.

Similar miniaturizations of all farm machines are needed. Some are already available, particularly tractors and related equipment. But work is needed to develop miniaturized harvesting equipment oriented toward making individual farm workers able to compete with large-scale machines.

4. Biological insect control. Much good work has already been done toward finding natural substitutes for toxic chemical pesticides, thanks to both the ecology movement and the realization some years back that pesticides are too expensive and have a limited useful life because of the build-up of insect resistance.

Increased scientific efforts in the biological control area are necessary. Of great interest are recent discoveries indicating that plants, animals and insects (and perhaps even man) are tied together in a chemical communication network. The active agents of this network are pheremones, essentially airborne or waterborne hormones. Pheremones provide the answer to many questions that have puzzled both biologists and farmers, and point toward new culture methods that eliminate toxic risks and lower costs of production. However, chemical pesticides cover up or interfere with the pheremone network, so the system of natural food production is not always compatible with partial use of chemicals, as in integrated control.

5. Educational technology. Thorough studies should be made of all ways in which both city and farm people could be taught to appreciate the virtues of small-scale production. Present education practices are directed toward creating agricultural specialists, or people motivated toward working in agribusiness operations.

6. Marketing techniques. Here is an area of great potential benefit for the small-farm movement. Intensive scientific and business efforts should be directed toward perfecting methods of getting fresh, relatively unprocessed food quickly and cheaply from farm to consumer. Cooperatives can be of help. So can improved packaging and shipping techniques.

On a recent visit to the U.S. Department of Agriculture, an editorial associate of mine requested that the USDA set up an "organic farming office" that would distribute useful information about managing a small farm by organic methods—for example, the cheapest ways to spread manure over fields; mechanical ways to control weeds; biological insect controls for the small farm; resistant varieties, and other subjects which the USDA obviously knows much about, but which farmers are not being informed about regularly by extension agents. Even a one-man office would be a start toward recognition by the USDA and land grant colleges that organic farmers are, in fact, a legitimate constituency to serve.

The request was turned down however, since—in the opinion of the USDA official—the Department already served not only all *farmers* but all *Americans*. The USDA and the land grant college complex have something for everyone, his reply went on, including organic farmers.

But over the years, everyone has come to be spelled with a capital E, and USDA policies reflect the recognition that agriculture is a Business also spelled with a big B. Evidently in the millions of dollars spent annually, there isn't much money or time left over to aid the family farmer—and certainly not the organic family farmer.

In early March, 1972 the Senate Subcommittee on Monopoly chaired by Senator Nelson held hearings on the role of giant corporations in American and world economics. Specifically, the witnesses tore into corporate secrecy and agribusiness. Repeatedly, the efficiency of the family farm was documented in every phase of food production and land management—but marketing.

However, it seems as if the organic market is emerging as a model for effective marketing by family farmers. It is becoming more widely recognized as just such an authentic model by consumers and even by some state officials. A recent editorial in *The Washington Post* confirms the recognition of organic as an alternative route:

> It is news to no one that a high tonnage of the food eaten every day by Americans is worthless, tasteless, contrived and can occasionally be actually dangerous to health. The production of all this junk food is a major U.S. industry. . . In many cases, the consumers who are rejecting it are turning to what are called organic foods . . . Though gimmickry and artificiality may one day become as much a part of the organic industry as it is now a part of the commercial food business, there is one built-in check. The shopper at the organic store is there precisely because he is suspicious of supermarket food. He is wary of the synthetic; he may or may not be a faddist lost in imaginary gardens of sesame seeds, but he also has a sense of the genuine.

There remains a very real danger that major companies will try to undermine the effectiveness of the organic alternative.

A continuation of present land grant college actions and philosophy will insure that there is no alternative to the destructive course of U.S. agriculture. Farms will get fewer and fewer, and farming profits will go to bigger and bigger conglomerates. More and more people—who want to remain on the land—will find their own tax dollars used to fight against the very agricultural alternative they are trying to create.

Right now, a sizable number of American consumers are paying a subsidy for foods grown by organic farmers. When you think about it, these Americans are being taxed twice in effect—first, all their regular tax dollars go through government channels to support and perpetuate chemicalized, agribusiness food production. Second, they are paying—voluntarily, I admit—an additional subsidy to encourage farmers to change away from methods which their official tax dollars support.

Existing efforts of land grant colleges are clearly not enough. Constructive programs will only develop when land grant college advisory committees and policies aggressively seek to develop the ways and means to help solve the problems plaguing family farmers and the people in rural communities. Half-hearted efforts—as we have seen in the past —get us nowhere. We need people in official capacities in the USDA and land grant college to say: "I am ready and able to support specific research and programs which will help more people make a better living on the farm. . . . I am ready and able to support specific alternatives to our present agricultural system."

This does not mean a condemnation of everything now going on in the agricultural system. This does not mean to be a call to stop all projects and issue statements like "We can do it, but you must pick which half of the U.S. will starve to death."

All I am saying is that those who seek change should have official recognition . . . should have a substantial amount of the dollars now being expended to support constructive change . . . and that people in high places should not be so quick to condemn those who would alter the agricultural status quo.

Through my involvement with *Organic Gardening and Farming* magazine, I am most familiar with the agricultural alternatives offered by the organic method. This is only one of the terms and forces now developing. I am sure that other labels and other terms will develop.

58

But we are witnessing a very vital development taking place around the identifying label offered by the word "organic." It has come to stand for an attitude that looks upon smallness as a virtue. In an era when most city people have grown up without any personal communication with the producer of their foods, the organic route is clearly different. Suddenly, the consumer can identify the farmer, and the farmer can identify the personal needs of the consumer. No longer is the supermarket clerk or the television commercial the most vivid contact. Suddenly, the organic family farmer replaces the Jolly Green Giant.

The word "organic" is helping city people to understand farming problems. The word is helping to forge an alliance between farmers and consumers. A great part of our present problems in society is due to programs that have actually built walls around farms and cities—programs that have isolated one segment of our society from another. This separation means that representatives of city voters vote against farm-oriented programs, and vice versa. Wouldn't it be great if more programs and alternatives stressed the common benefits to both city and rural people?

I ask the Senators exploring the land grant college system to look upon the needs of organic farmers and organic food customers as a step toward developing future programs which that system could develop. These people want to tear down the barriers to communication. These people are against the present trend where farmers go indifferently in one direction, while consumers go in the other—each blaming the other for their respective troubles.

Out of these hearings should come a clear recognition that the purpose of the land grant college system is not to create only one single agricultural system that helps only those who are big enough to plug into. Diversity is a healthy characteristic of all environments. And we need a land grant college system that thrives on diversity.

In May, 1972, our company sponsored a National Conference on Organic Farming and Composting to report on how cities are using—and can use—organic wastes like sewage sludge to build soils, and how the world needed an agricultural system that makes use of those organic wastes.

We believe organic farming can provide such an agricultural system. Organic farming can provide high-quality food to consumers in nearby cities, farms that can provide jobs, farms that can be both economically-and environmentally-sound. And farms that can use composted city wastes to build humus into soils.

The Conference brought together qualified experts in solid waste management and public health, but we were unable to secure a single representative from the land grant colleges to present a report on those topics. As has been the case with the development of organic agriculture in this country, it continues to be the responsibility of proponents of an organic agriculture to be their own researchers, their own experimenters, their own extension services—while the tax-supported research into agribusiness goes on and on.

What is perhaps saddest of all is that this perversion of agriculture has probably come about from what starts out to be the noblest of motives—the best and most food at the cheapest price. With such motives, the egg industry has been revolutionized. And so has the beef industry. And so has every single crop. But has the Maine potato farmer been helped? Or the Wisconsin dairymen? Or the New Jersey egg farmer? Or the California truck gardener? Undoubtedly some have been helped—the relatively few who have survived, perhaps. But isn't it time to begin new policies—new programs that will specifically aid small farmers? I think so!

For years, we as publishers have reported on developments in organic agriculture. We have in a modest sense acted as a kind of extension service for organic growing methods, relaying information. But the need now is too great and the hardships too severe to continue as we have in the past—hoping for a recommendation here and a bit of advice there. We believe it is time for the land grant colleges to give organic agriculture all the positive help it can.

It is time to stop playing games—to dismiss as insignificant alternatives that are already helping black sharecroppers in Virginia and Georgia to earn a decent profit by supplying organically-grown cucumbers to urban markets.

It is time to make the most of such alternatives—and stop treating a genuine consumer demand for quality food as fraudulent—or a genuine back-to-the-land movement by people of all ages as merely a fad. What must the American people do to convince a Department of Agriculture official that their goals are not satisfied by continued all-out drives for efficiency and specialization?

A New Homestead Act

Peter Barnes

Chapter 6

I'm a city-based writer, not a farmer, and a consumer of non-organic as well as organic foods. For a long time, I instinctively thought that there was no other way to live except in cities, stacked up in concrete boxes alongside thousands of other people who never speak to each other. Eventually it began to dawn on me that there might be other ways and places to live, closer to the land, more in harmony with nature and with other human beings. I began to do research and write articles about farming and rural life. Much to my dismay I discovered that many shocking developments were taking place.

The first thing that struck me was that, despite the congestion and decay and pollution in most American cities, people are still leaving the land and flocking to the cities in droves. This is not because they want to leave the land but because they are forced to. Every week about two thousand family farms go out of business. Usually these are farms that have been owned by oldtimers who've been trying to scratch out a living for years, are heavily in debt, and finally throw in the towel. Thousands of farmworkers and sharecroppers also leave the land each year because they've been replaced by machines developed by the big agricultural universities. Everybody complains when a big company like Lockheed is forced to lay off workers, but nobody gets excited when farmers or farmworkers lose their jobs. That's progress, the experts say. It's also slums, rising welfare rolls, and a lot of human misery.

A second thing I saw happening in rural areas was a steady take-over by giant corporations. In some parts of the country, particularly the West and the South, this has been going on for some time. These areas have long been characterized by large landholdings—the big plantations in the South, the enormous ranches of the West. In recent years, these large landholdings have come under increasing control of corporations based in New York, San Francisco, Houston and other cities.

It's very difficult for even the most efficient family farmer to compete against a giant conglomerate such as Tenneco, a multi-billion dollar enterprise that's near the top of *Fortune's* list of American corporations. Tenneco not only owns or leases more than a million acres of land in the West, it also sells oil and natural gas, makes farm machinery and fertilizers, builds ships for the Defense Department, packs and distributes its own foods, and spends millions to advertise its brand name. Tenneco executives say they want to build a totally integrated food system that is Tenneco-owned "from seedling to supermarket"—and they're well on their way to achieving it.

Other giant corporations are playing the same game. Among the blue-chips that have lately plunged into agriculture (often as a way-station on the road to land development) are Dow Chemical, Monsanto, Union Carbide, Kaiser Aluminum, Aetna Life, American Cyanamid, Goodyear, W. R. Grace, Getty Oil, Purex and Coca-Cola. You'll notice that a lot of these are oil and chemical companies. Along with a few dozen timber, coal and railroad companies, these big conglomerates are now the effective rulers of rural America.

What are the consequences of the corporate invasion of agriculture? For one thing, small-town businesses are dying because city-based corporations don't make their purchases locally. In addition, farmers—if they're not forced off the land—quickly lose their independence. They become mere cogs in a corporate-dominated food production system. For example, Harrison Wellford, an associate of Ralph Nader, has described in a recent book (*Sowing the Wind*) how once-independent poultry farmers have become virtual peons of Ralston-Purina and other agribusiness corporations. In some cases their expenses actually exceed what the corporations pay them, so they wind up working for *minus* 30 cents an hour or even less!

The same sort of thing has been happening in the cattle industry. As Victor Ray of the National Farmers Union has pointed out, cattle feeding in Colorado has been taken away from family farmers and is now a corporate operation, controlled by packers. What it all boils down to is this: the same corporations that produce chemicals and put additives in our food are now squeezing out the small independent farmer. The corporations claim that this will benefit the consumer, but I haven't noticed supermarket prices getting any lower, or the quality of food getting any higher.

There's another consequence of big-scale corporate farming—the loss by rural citizens of control over their own communities. I remember driving through the town of Mendota, California, not long ago and talking with Jack Molsbergen, a local realtor. Molsbergen told me how the people of Mendota had wanted to construct a local hospital, since the nearest one is 40 miles away. But three giant corporations that own more than half the land around Mendota opposed the hospital plan and killed it. "Why would a corporate executive who lives in Houston give a hoot about a hospital in Mendota?" Molsbergen asked.

Another alarming development in rural America is the slow but steady deterioration of the environment—the spread of plastic, suburban-type developments over what was once prime cropland, and the poisoning of the soil through excessive use of inorganic chemicals. Family farmers, of course, are frequent abusers of fertilizers and pesticides, but the biggest offenders are giant corporations— spray now and pay later seems to be their motto. The corporations are also most actively engaged in land speculation and development, activities which drive up the price of land and make it even harder for new farmers to get started, and for old farmers to pay their taxes.

Perhaps some corporate executives do care about what they are doing to the land, but I doubt that many do, or that they can really love the land the way a farmer does who lives and works on it every day. I suspect that most corporate executives would agree with Simon Askin, a vice-president of Tenneco, who was quoted in the Los Angeles *Times* as saying, "We consider land as an inventory, but we're all for growing things on it while we wait for price appreciation or development. Argiculture pays the taxes plus a little."

The worst part about what's happening to farms and to rural areas in America is that none of it is accidental. The exodus of people from the land to the cities, the domination of rural communities by absentee corporations, the subdivision of good cropland and the massive use of inorganic chemicals—all these things are considered *desirable* by the U.S. Department of Agriculture and are promoted by a wide variety of government policies.

For anyone who wants a good understanding of where the U.S. government thinks we ought to be heading, I strongly suggest reading the February, 1970 issue of *National Geographic*. Here are stunning photographs of an egg factory near Los Angeles where two million caged Leghorns gobble 250 tons of feed and lay one million eggs each day; a cattle metropolis in Colorado where 100,000 steers fatten on formulas prescribed by computer; a $23,000 tomato harvesting machine, developed by the University of California, that snaps up hard-skinned tomatoes especially bred for mechanical picking and packing.

The most frightening picture in the *National Geographic* article is an artist's depiction, under USDA guidance, of a typical American farm of the future. The farm—if that is what it can be called—is more than ten miles long and several miles wide. Obviously, only a giant corporation could own it. All operations are monitored from a bubble-top control tower by one man. An enormous remote-control tiller glides across a ten-mile long wheat field that has been leveled with nuclear explosives. Overhead, a jet-powered helicopter sprays pesticides. Cattle are housed in skyscraper feedlots. Underneath the soil are sensors which find out when crops need water, and automatic irrigation systems that bring it to them. Far in the background is a city where presumably the people who once lived on the land now reside. If they are employed at all, it is probably on assembly lines or at dull, meaningless clerical jobs, in buildings whose windows are permanently sealed shut. For dinner they probably eat pre-cooked, over-priced foods laced with chemicals.

I should repeat that this vision of the future is not the fantasy of some wild-eyed science fiction writer but the desire of high U.S. government officials and powerful corporate executives. It is, moreover, the kind of future toward which existing policies are leading.

Tax laws, for example, strongly favor the invasion of agriculture by large corporations and investment syndicates. They can use losses from farming to offset profits from other sources, a luxury that the genuine farmer does not enjoy. Or they can use investments in land to transform non-farm profits into capital gains, a form of income that is taxed at half the normal rate. Several of the Wall Street sodbusters also receive tax breaks of a different sort—they are oil companies that benefit from the oil depletion allowance and other generous loopholes. For example, our old friend Tenneco not only paid no federal income tax in 1970, it actually wound up with the federal government owing it $20 million. Not bad for a company with profits that year of $100 million!

Then there are the federal crop subsidies (often for *not* growing crops) that reward big corporations far more than the small family farmer, the farmworker or the sharecropper. In 1970, the J. G. Boswell Company of California received federal subsidies totalling $4.4 million; Tenneco got $1.5 million, and the list goes on like that. Now there's a $55,000 ceiling on the subsidies that any one farmer can legally receive, but the big boys know how to get around such limitations.

Another policy that favors the giant corporations: dams and irrigation canals that carry year-round water to private landholdings, regardless of size. When Congress first authorized the reclamation program in 1902, it specifically stated that no water would be delivered to farms exceeding 160 acres, or to farms whose owners did not live on or near the land. The decision to limit the benefits of federally-subsidized irrigation projects to small owner-occupied farms was a wise and farsighted one, but it has been all but ignored by the Interior Department.

Fortunately, not everyone accepts the government's idea that big-scale corporate farming is the wave of the future. Most farmworkers and sharecroppers that I have talked to want to stay on the land if they can earn a decent living.

Several have started cooperative farms to show that there is an alternative to welfare and city slums.

Family farmers have also been raising hell about the corporate invasion of agriculture—though not as much hell as they should raise. Organic farmers in particular have demonstrated that there is an alternative to large-scale "factories in the field" that is ecologically sound and economically viable.

But the sad fact is that the tide is far from being turned, and unless we turn it soon there will be no way to stop it. Organic farms and cooperatives provide a model of what the future can be. But we need more than models if we are to stop the corporate takeover. We need new laws—laws with teeth in them that will give the land back to the people who live and work on it.

What kind of laws do we need? I don't profess to have all the answers, but I can see several things that badly need to be done.

First, we need new tax laws that don't reward the speculators, tax-loss farmers and giant conglomerates. Perhaps we should eliminate the preferential treatment for capital gains, the oil depletion allowance and other major loopholes. Perhaps we should have progressive property tax that falls most heavily on large landowners, and lightens the burden on owner-occupied homes and small farms.

Second, we need to end the subsidies to big corporations. If there must be subsidies (and there probably should be to stabilize farm income), they should go to the working farmers who need them most.

Third, we need laws that restrict the use of toxic chemicals and set standards for organic products.

Fourth, we need laws to strengthen the 160-acre limitation and the residency requirement in government irrigation projects. Senator Fred Harris of Oklahoma has introduced a measure that would enable the federal government to purchase large landholdings in reclamation areas and re-sell them in family-sized parcels—as Congress originally intended. It's a good bill that ought to be passed.

Fifth, we need new anti-trust laws to get the corporations out of agriculture. One such law would be the Family Farm

67

Act, introduced by Senator Gaylord Nelson and representatives of both political parties.

Sixth, we need a Homestead Act for the Twentieth and Twenty-first centuries that will put people back on land that belongs to them. The government can't give out free land any more, but it *can* purchase and re-sell lands in rural areas, just as it does in urban areas (where it's called "urban renewal"). It can also provide for re-sale on terms that a poor person can afford. This might be done the way some colleges make loans for tuition: the student agrees to pay back a fixed percentage of his future earnings. If he gets rich, he pays back a lot; if not, he isn't overburdened with debt.

Of course, it's easy to propose new laws, but a lot harder to get them passed. The big agribusiness corporations are very powerful politically. They have friends throughout Congress and the executive branch—for example, the new Secretary of Agriculture, Earl Butz. They contribute heavily to political campaigns. They hire lobbyists who are expert in the ways of Congress and the state legislatures. Only a broadly based and well-coordinated effort of concerned citizens throughout the country stands a chance of overcoming that kind of power.

Recently I joined a group that is trying to organize just such a citizens' effort—the National Coalition for Land Reform, with head offices at 126 Hyde Street, San Francisco, California 94102. The Coalition is made up of people from all walks of life—family farmers, farm workers, sharecroppers, organic growers and consumers, environmentalists, clergymen, labor leaders, students—anyone interested in joining the fight.

Other organizations are engaged in the effort in different ways: the National Sharecroppers Fund, the National Farmers Organization, the National Farmers Union, the United Farmworkers Organizing Committee, the Agribusiness Accountability Project and the Sierra Club, to name a few.

The important thing is that we all get involved in one way or another. If we don't we will wake up one morning and find the American countryside looking like it does in the *National Geographic* picture, and be unable to do anything about it.

A New Policy Direction for American Agriculture

Marion Clawson

Chapter 7   The American people face major policy decisions about the nature of rural America in the next quarter century. Several basic agricultural adjustment laws expire soon, and they must be renewed, modified, or abandoned. The difficulties inherent in our capacity to produce surplus agricultural commodities, the inelastic demand for these commodities, and rising output due to increased technology persist as they have for several decades. Past efforts to adjust to these difficulties have not been successful, and further adjustments are necessary.

How far has agriculture come the past 25 years? Where will it be 25 years from now? More importantly, where do we want it to be, and how can we get it there?

The term "agriculture" encompasses far more than production and marketing. All aspects of social and community life in farming areas are as much a part of agriculture as the specific farm problems that so often are the sole concern of agricultural policy discussions. Farm people have always been concerned with the quality of rural life. Farming is a business, but life is more than a business.

Evolution of Farm Programs

Federal programs to improve agricultural income began in the Farm Board days of the late 1920s and early 1930s. They expanded and changed greatly in the early days of the New Deal. The country then was in the tight grip of an economic depression, the severity of which is beyond the understanding of millions of young people today. A fourth or more of the total labor force was unemployed; people eked out a living selling apples on street corners; "Hoovervilles" sprang up as displaced persons sought some form of meager shelter on the edges of cities and towns; and farm prices fell to unbelievably low levels. In some places a bushel of oats would not buy a postage stamp. Corn was used for fuel in kitchen stoves because coal was too costly. Farm mortgage holders foreclosed by the thousands. At public auctions, foreclosed farms were bid in for one cent by farmers who would not tolerate displacement of established farmers.

In a desperate effort to redress the disparity between agricultural production and effective market demand, agricultural programs of the New Deal instituted various forms of production control, price support, and storage of surpluses.

Through Democratic and Republican administrations for nearly 40 years, this trinity of production control (or supply management), price support, and surplus storage has remained central to all agricultural programs. Some changes have occurred, of course, but it is fair to characterize these changes as a warming over, not as the cooking of a new dish. Through depression and boom, through war and peace, the same basic programs have continued.

These programs have largely failed. They clearly have not cured our strong tendency to overproduce and thereby lower prices. The demand for their continuance evidences this. Furthermore, they have produced serious side effects. They have not aided, and may have worsened, the malaise that effects rural communities, and they have become costly at a time when competition for a piece of the federal budget is increasing. Truly basic changes in agriculture have proceeded more or less independently of them. It is time to consider a different approach, at least for the longer run.

Cropland Limitations

A basic ingredient in this continued agricultural adjustment effort has been the limitation on the use of cropland. From the original Agricultural Adjustment Act, through the Soil Bank, to current programs, varying amounts of cropland have been taken out of production. The input of land as one factor of production has been reduced compared to what it otherwise would have been. No effort has been made to control other productive inputs: fertilizer use has increased, offsetting to a considerable extent the reduction in cropland area; improved technologies of all kinds have been developed and pushed by public agencies and private firms; productive capital has been provided, often at a subsidy; and no effort has been made to reduce labor input, although economic forces have substantially reduced labor use. Payments to farmers have been based on crop acreage allotments or other devices geared to cropland area. Ownership or control over land has been the sole means of securing federal subsidies or payments.

Federal programs as a whole have been inconsistent even on this matter of land area. While one branch of the U. S. Department of Agriculture has taken cropland out of production, other branches have helped farmers make more intensive use of the land left in production, and USDA and other federal departments have heavily subsidized drainage, irrigation, and flood protection to increase land productivity.

One major effect of this emphasis on land as a factor of production has been a persistent increase in farm land prices. The total value of all farm real estate in the United States has more than doubled since 1954. In the same period net income to agriculture as a whole has remained nearly constant.

Various factors are responsible for this marked price increase, but direct payments from the federal government to landowners, based on land area and crop production history, have surely been a major factor. The ability to collect a tobacco, wheat, cotton, corn, or other income or price support on land formerly used to produce these crops has been translated into higher land prices.

Far more serious has been the creation of an artificial land scarcity by federal agricultural programs. At present, more than 50 million acres of cropland are held out of production by USDA farm programs. A reduction in cropland by this amount has had a major effect on land prices. If the price of cropland is highly inelastic, as seems probable, a reduction of about an eighth in land supply would result in an increase in land price of a fourth or a half.

72

The conventional wisdom of a generation ago was that farmers gained by rising land prices. Even then it was conceded that the young, beginning farmer had to pay more for his farm, creating a hardship on him. But established farmers were supposed to gain. Only in this way, it was argued, could farmers accumulate savings for their old age.

This reasoning implicitly embraced a theory of a static agriculture to a degree that its advocates did not realize. If farm size remained constant through a farmer's lifetime, he could buy his farm cheaper when he was young and sell it for more when he was old, and thus gain more.

During the past generation, however, average farm size in the United States has doubled. In recent years about half of all purchases of farmland have been by established farmers to increase the size of their farm. What they may have gained in increased prices of land they previously owned, they have lost in increased prices of land they have bought.

The total value of farm real estate in the United States today is so high that an interest return at competitive interest rates leaves nothing for wages to the farmer. Many a farmer can earn interest on his farm investment only if he is willing to forego all wages from his work. Conversely, he is able to earn wages only if he is willing to forego interest on his farm investment.

Agriculture is rapidly becoming a high-investment, low-wage industry. This is unhealthy.

Federal agricultural programs of the past generation have given only incidental attention to the human problems of farm people. The farm owner with no debt on his land has gained from higher prices of agricultural commodities, if his volume has been large enough. The small farmer, whose basic problem is inadequate farm size, or the farm laborer or tenant without even a minimum farm has benefited little or not at all from federal farm programs. Programs were intended only to benefit the larger commercial farms. Then Under Secretary of Agriculture John Schnittker stirred up a hornet's nest in 1968 when he said just this.

The old Farm Security Administration sought to aid tenant farmers, sharecroppers, and small farmers generally. It has long since been emasculated. Ensuing farm programs have been directed to farm property and the owners of such property while other rural people have been left in the cold.

In spite of the magnitude and persistence of federal agricultural programs, the basic changes in agriculture in the past generation have proceeded largely independently. There has been a revolution in agricultural production—output has greatly increased; cropland area has shrunk; labor input has decreased by more than half; and total capital input, although changed somewhat in form, has only modestly increased.

The pace of change in American agriculture during this past generation has been far swifter than in industry in the same period or than in industry during the Industrial Revolution 100 to 200 years ago. Farm numbers have been cut in half, primarily because young men have been so repelled by the outlook for low incomes that they refuse to enter farming. Farming has become an old man's occupation. Many rural communities are in serious decline. In all of this, federal agricultural programs have been unimportant, often impeding adjustments rather than facilitating them, rarely helpful to the individuals concerned with easing the pain of their personal adjustments.

Some groups have been injured by agricultural programs. A generation ago, Negroes were largely rural; thousands were sharecroppers. Today they have almost disappeared from agriculture, except as casual hired laborers and these only in limited numbers. The old sharecropping is hardly an ideal form of life or business for anyone in a modern world, and it can be argued that the Negro sharecropper of pre-World War II days was doomed economically. But the choice to stop farming frequently was not his but his landlord's, and the landlord was motivated in part by federal agricultural programs as well as by mechanization and other technological changes.

Today, there are about three million farms in the United States. At least two-thirds of these are unneeded in the long run. Over 50 million acres of cropland are currently lying idle as a result of government programs. Over 300 counties in the United States have lost so many young people that the number of young people of reproductive ages remaining is so small that the number of babies born each year is less than the number of old folks that die. Rural towns that depend on farming for customers are disappearing.

The likely economic and social trends of the future will intensify many of these problems; some major economic, social, and political adjustments in rural America seem inevitable.

Future Farm Programs

This poses these questions: What do we really want of our farming and rural areas? What kind of a business do we want farming to be? What kind of rural communities do we want?

The old ways of farming are disappearing under the impact of science and technology. Old rural communities are disintegrating for the same basic reasons. What do we wish to replace each?

I propose a more considered, rational, purposeful set of public programs for agriculture.

Americans are not likely to continue to appropriate $6 to $7 billion annually for agriculture in the face of a constantly shrinking agricultural population and steadily rising competition for federal appropriations for other domestic and international programs. More than this, I doubt if we should appropriate such sums for agriculture. While I sympathize with farmers and rural people, it seems that the rest of use have a right to ask what we are buying with so much money. There is no evidence that costs are decreasing or that a permanent solution to agricultural problems has been achieved.

The most discussed new approach to the farm problem concerns federal purchase or long-term leasing of cropland, largely in whole farms, as a device to bring supplies of farm commodities in line with demands, at "fair" prices. Proposals have been advanced to buy or lease more than 50 million acres of cropland in the next few years.

This approach is just more of the same. It would be costly and disruptive to rural areas. It would leave the basic adjustments in farm size, farm production, labor force, rural community, and the rest to be worked out by the farm and rural people concerned. It is a tested approach—tested and failed.

A Proposed Policy Direction

A critic has some obligation to say what he would do to better the program or proposal he is criticizing. This is not the time or the place to detail a better farm program. In any case, there is more than one way of achieving broad objectives, and I do not want arguments over details to obscure the broad lines of a new policy direction.

In their starkest outlines, here are my suggestions for a different policy direction for agriculture in the next quarter century:

1. Complete the shift away from price supports (with attendant problems of surplus storage or supply management) to income support. This has been done for wool, partially done for cotton, and could easily be done for other commodities. If farm incomes are below a socially acceptable level, raise them by public action. Free prices of agricultural commodities from their income enhancement role, and let them serve as resource and consumption allocators. Artificially manipulated prices seriously distort both production and consumption.

2. Allocate the income support payment to the farmer as an individual, not as a landowner. It might be necessary to divide the income payment between landlord and tenant, but each should receive it as a person, irrespective of landownership. Provide no income support payments to new farmers, even to sons of present farmers. Income support should be a one-generation phenomenon, with an end in sight. Under conditions outlined below, allow the farmer to continue to draw his income support even when he quits farming. Instead of pushing up land prices, encountering the problems this entails for adjustments in farm size and income allocation, let the market for land operate unre-

stricted. Extend personal aid to present farmers but not to future ones.

3. Encourage older farmers on inadequate farms to retire early. As a possibility, grant farmers 55 years and older the social security payments they would earn at age 70, assuming they had the maximum coverable income each year. In return, require these farmers to sell their farms to the federal government for one-half the estimated market price. Since the farmer would gain a substantial subsidy through his retirement income, he should be required to make some sacrifice to obtain it. Selling his farm at half the market price would be one form of sacrifice. Early retirement would obviously be voluntary, and no farmer would choose it unless his expected retirement gain would more than offset any immediate loss in the market value of his farm.

4. Enable any farmer to take his farm income support payment with him for life, even though he no longer farmed or lived on the farm, subject to the same requirement of selling his farm to the government at half its market price. His successor would not be eligible for income support from that farm. This proposal would appeal to the marginal farmer whose income elsewhere would be nearly as good as his income on the farm. Sometimes this would be the farmer on a marginal farm, sometimes not. It would often appeal to smaller and younger farmers who could adjust to other jobs and living in town. The previous proposal of early retirement would often appeal to older farmers on small farms.

The costs of the two foregoing proposals could be higher, the same, or less than the present cost of price support, depending on the level of income maintenance. There is nothing about the form of the payment that determines its cost, and arguments that farm income maintenance would be more costly than price support programs are nonsense. Farm income support tied to the individual would have the major advantage that an end would be in sight. Regardless of how generous the payments might be, they would end when the recipient died. Experience over the past 35 years has demonstrated clearly that there is no end to price support. It is as costly today, or more so, than it was two decades or more ago, and no one predicts its end as long as present programs continue.

5. As a result of the foregoing, the federal government would acquire title to considerable tracts of farmland. To this extent, this proposal includes federal land purchase, as do other current proposals. But the role of land purchase differs. I would make it incidental to human adjustment rather than the basic agricultural adjustment to which people must conform. This proposal includes voluntary participation on the part of farmers but compulsory sale of land if they are to receive other benefits. Sale of land to the government under these conditions would not bid up land prices. It might even tend to push them down.

6. The use the federal government made of the land it bought in this way would be a major part of the program. Lands needed for public purposes, such as parks or wildlife refuges, would be retained in public ownership. Most would be sold to private individuals. First, however, lands should be "reconditioned" if necessary. A substantial part of the soil conservation program could be shifted to this end. Some lands might be sold subject to land use restrictions. For instance, some Great Plains land could be sold for grazing but not for cropping. Some lands might be sold to cooperatives or to public-private land corporations, in each case for private grazing use. Much of the land could be sold on the open market. A requirement might be attached limiting purchase to someone already owning farmland, thus insuring that the land be used only for enlarging farms. I see no reason why much of this land should not be owned and used for agriculture, since production adjustments would be secured in ways other than by artificial limits on land use. Sale by competitive bidding would insure the government of getting the best return for the land and, at the same time, exert some stabilizing or downward pressure on land prices. Instead of creating an artificial scarcity of farmland by holding large acreages out of production, this proposal would free-up the land market by bringing additional supplies of land into the market.

7. The federal government would invite states, counties, and small towns to launch a cooperative endeavor to

strengthen some small towns while helping people in other small towns relocate in more viable communities. By providing better public services in some larger but still moderately sized towns and by concentrating government functions there, these areas would be greatly strengthened and could provide much better social and economic services. Some small towns and rural areas are clearly on the downgrade now. Instead of trying futilely to prop them up, people could be relocated in larger and more viable cities. Early retirement, income aids, purchase of property, and the other devices suggested for farmers could be extended to residents of such areas. In addition, direct help could be given to school districts, other special districts, and counties to encourage consolidation into more efficient governmental units. A program of this kind would require research, and hopefully, agricultural colleges and other institutions might rise to the challenge of this research better than they have in the past. Likewise, assistance might be extended in transportation planning to reduce the excessive road mileage that exists in many rural areas. Farmers are increasingly shifting their residences from the farm to the town, and research might well indicate ways in which this process could be aided. A highly flexible approach, varying to meet circumstances of different areas and geared to take advantage of natural changes arising with time, could, in a decade or two, greatly strengthen many rural communities.

The foregoing is a "bare bones" outline. Details and specifics would have to be developed if there were agreement on general approaches. The purpose of the proposal is to develop an economically healthy agricultural industry, as free of government controls and regulations as possible and free of large numbers of inefficient, small farms. Farmers would be helped to find satisfactory jobs and satisfactory lives on the farms or in towns rather than be subsidized to hold land out of use. The emphasis would be on rural people, not on farm property and commodity prices. There would obviously be many problems to work out, but an end would be in sight.

No specific program for adjusting agricultural output is included. There would be a transition period, but in the end, output, prices, and land use would be adjusted by the market. Interference with a free market would be limited to the help extended to farmers on small, outmoded farms to get a better job and a better life. There would be more of a floor under personal farm income than now and far less interference with agricultural commodity prices.

Role of Conservation Programs

One concluding comment is in order. Over the past 25 years, soil conservationists have seen "conservation" used as a peg for many agricultural programs, some of which were far from conservation. Conservation is a term with wide public acceptance, and anything called conservation almost automatically acquires considerable public support. Soil and water conservationists must consider how far and in what ways they have been used as a front for agricultural adjustment programs with little, if any, real conservation content. One should not assume that all such use or misuse of conservation is ended. Agricultural adjustment programs must stand squarely on their own foundation, as should conservation programs.

The Importance of Cities to Rural Living

Jane Jacobs

One of many surprises I found in the course of this work (*The Economy of Cities*) was especially unsettling because it ran counter to so much I had always taken for granted. Superficially, it seemed to run counter to common sense and yet there it was: work that we usually consider rural has originated not in the countryside, but in cities. Current theory in many fields—economics, history, anthropology—assumes that cities are built upon a rural economic base. If my observations and reasoning are correct, the reverse is true: that is, rural economics, including agricultural work, are directly built upon city economics and city work.

Chapter 8

We are all well aware, from the history of science, that ideas universally believed are not necessarily true. We are also aware that it is only after the untruth of such ideas has been exposed that it becomes apparent how pervasive and insidious their influence has been.

In just such ways, I think, our understanding of cities, and also of economic development generally, has been distorted by the dogma of agricultural primacy. I plan to argue that this dogma is as quaint as the theory of spontaneous generation, being a vestige of pre-Darwinian intellectual history that has hung on past its time.

Early in the eighteenth century, a great improvement was made in crop rotation, a change so important that it is at the heart of what is called "the eighteenth-century agricultural revolution." In the former fallow year, crops not previously employed in the rural farming of Europe were planted: alfalfa, clover, and another fodder crop called sainfoin. The fodder crops did more than "give the land a rest." They replaced nitrogen used up by grain and at the same time supported cattle. The livestock provided nitrogen-rich manure. Fertility of the cropland and numbers of farm animals both increased at an extraordinary rate, making pos-

81

sible the abrupt European population increases that so alarmed Malthus.

Where did rural Europe get the fodder crops, along with the practice of fitting them into the rotation in place of the fallow year? Duby and Mandrou say the fodders were being grown in the city gardens of France for at least a century before they were adopted into rural farming, and that they were also grown in nearby fields to feed city draft animals. As in the case of the twelfth-century rotation, the new agriculture spread first near cities and along the trade routes, and it was adopted last in the rural areas most distant from cities and least touched by their trade and goods.

The idea that agriculture itself may have originated in cities, the thought to which I have been leading, may seem radical and disturbing. And yet even in our own time, agricultural practices do emerge from cities. A modern instance has been the American practice of fattening beef on corn before slaughter, the practice that has given us the corn-fed steak. This "farm work" did not begin on farms or cattle ranches, but in the city stockyards of Kansas City and Chicago. It was a forerunner of such present farm work. The fattening pens are all but gone from the cities now because the work has been transplanted from cities to the rural world.

Both in the past and today, then, the separation commonly made, dividing city commerce and industry from rural agriculture, is artificial and imaginary. The two do not come down two different lines of descent. Rural work—whether that work is manufacturing brassieres or growing food—is city work transplanted.

If my reasoning is correct, it was not agriculture, for all its importance, that was the salient invention, or occurrence if you will, of the Neolithic Age. Rather it was the fact of sustained, interdependent, creative city economics that made possible many new kinds of work, agriculture among them.

For instance, if fatal misfortune dealt either by men or by nature befell a parent city, then its farming villages—if they managed to survive the disaster—would be cast loose with their incomplete fragments of a rounded economic life. These orphaned villages would, of course, continue to specialize—do the work they could do—but now only for their own subsistence. They would not develop further because there would be no parent city economy from which they might receive new technology. Again and again during prehistoric times, villages must have been orphaned by the destruction of cities.

Just as no real separation exists in the actual world between city-created work and rural work, so there is no real separation between "city consumption" and "rural production." Rural production is literally the creation of city consumption. That is to say, city economies invent the things that are to become city imports from the rural world, and then they reinvent the rural world so it can supply those imports. This, as far as I can see, is the only way in which rural economies develop at all, the dogma of agricultural primacy notwithstanding.

Solutions to most of the practical problems of cities begin humbly. When humble people, doing lowly work, are not also solving problems, nobody is apt to solve humble problems.

Cities as Mines

Let us now look a little into the future. If we observe the acute practical problems of cities in highly advanced economies today, we may be able to glimpse some of the forms economic growth could take in the highly advanced economies of the future—wherever such economies may prove to be. Waste disposal will do as an example, for in many different forms—air pollutants, water pollutants, garbage, trash, junk—wastes have created highly acute problems for large cities. They cause lesser problems, which are nevertheless chronic and unsolved, outside of cities.

Although the cities of the United States are making little or no progress in coping with wastes, hints and clues to solutions do appear. What they portend, I think, is not waste "disposal," but waste recycling. Odd little news items about wastes crop up.

Here and there, garbage is being processed into compost. The *New York Times*, which seems to employ someone deeply interested in garbage, has described a little factory in Brooklyn, New York (run by the proprietor and a part-time helper) that converts restaurant garbage into light-weight, pulverized, dehydrated garden compost. The income from the sale of the compost is clear profit; the proprietor of the plant pays his costs by means of the silver he retrieves from the garbage and sells back to the restaurants.

One glimpses how waste recycling can be made economically feasible even while it is still in a primitive and experimental state. Cities that take the lead in reclaiming their own wastes will have high rates of related development work; that is, many local firms will manufacture the necessary gathering and processing equipment and will export it to other cities and to towns.

A type of work that does not now exist is thus necessary: services that collect all wastes, not for shunting into incinerators or gulches, but for distributing to various primary specialists from whom the materials will go to converters or reusers. The comprehensive collecting services, as they develop into big businesses, will use many technical devices. They will install and service equipment for collecting sulfuric acid, soot, fly ash and other wastes in fuel stacks, including gases that, at present, cannot yet be trapped. They will supply and handle containers for containerized wastes and will install fixed equipment such as chutes, probably by employing subcontractors. *Who will develop the comprehensive collecting services?* My guess is that the work, when it does appear, will be added to janitorial contracting services—a kind of work itself that as yet hardly exists except for the benefit of relatively few institutional and other large clients, and is not notable for yielding development work. But in economies where people doing lowly work are not hampered from adding new work to old, we may expect that just such lowly occupations as janitorial work will be the footholds from which complex, prosperous, and economically important new industries develop.

The first priority in dealing with water-borne wastes, in view of the difficulty of mining them, is to keep them out of the water to begin with, if at all possible—to collect them in some other way. This is indeed possible with some water-borne wastes: those that are in the water only and solely because the water is a means of carrying them away from the point of production. Human excretions are in this category; to carry these wastes away by flowing water is extraordinarily primitive. It is amazing that we continue to use such old-fashioned makeshifts. Excrement in sewage complicates the handling of all city waste waters, including even the runoff from rainstorms, and exacerbates all the problems of public health connected with water pollution.

Developing economies are all too ruthless to nature, but their depredations do not compare in destructiveness to those of stagnating and stagnant economies where people exploit too narrow a range of resources too heavily and monotonously for too long, and also fail to add into their economies the new goods and services that can help repair their depredations.

The effects of economic stagnation upon nature are veiled when populations are so scanty and so primitive in their technologies that anything they do has relatively little effect upon the rest of the natural world. But once a society has developed its economy appreciably, and thus has increased its population appreciably too, any serious stagnation becomes appallingly destructive to the environment. Common sequels in the past have been deforestation, complete destruction of wild life, loss of soil fertility and lowering of water tables. In the United States, lack of progress in dealing with wastes, and overdependence on automobiles—both evidence of arrested development—are becoming very destructive of water, air and land. Poverty has no causes. Only prosperity has causes. Analogically, heat is a result of active processes; it has causes. But cold is not the result of any processes; it is only the absence of heat. Just so, the great cold of poverty and economic stagnation is merely the absence of economic development. It can be overcome only if the relevant economic processes are in motion. *These processes are all rooted, if I am correct, in the development work that goes on in impractical cities where one kind of work leads inefficiently to another.* Let us now get down to examining the movements that go on within the economies of cities—the little movements at the hubs that turn the great wheels of economic life.

Great surges of agricultural expansion coincide with great surges in city jobs—not great surges in rural jobs. Rural jobs, in fact, decline proportionately, and even absolutely, at precisely the times of great surges in a nation's agricultural production; of course, as we have seen, the workers who do remain in agriculture become more productive.

A recent experience of the Rockefellers in India illustrates this point. The Rockefellers, early in the 1960s, decided to build a factory in India to produce plastic intrauterine loops for birth control. At the same time they were undertaking to combat the Indian birth rate, they also wanted to curb the migration of rural Indians to cities. A way to do this, they thought, was to set an example of village industry, placing new industry in small settlements instead of cities. The location they chose for the factory, then, was a small town named Etawah in highly rural Uttar state. It seemed plausible that the factory could as well be located one place in India as another. The machinery had to be imported anyway and the loops were to be exported throughout India. The factory was to be small, for with modern machinery even a small factory could begin by turning out 14,000 loops a day. The work had been rationalized into simple, easily taught tasks; no pre-existing, trained labor pool was required. The problem of hooking up to electric power had been explored and judged feasible. Capital was sufficient, and the scheme enjoyed the cooperation of the government of Uttar.

But as soon as the project was started everything went wrong, culminating in what the *New York Times* called "a fiasco." No single problem seems to have been horrendous. Instead, endless small difficulties arose: delays in getting the right tools, in repairing things that broke, in correcting work that had not been done to specifications, in sending off for a bit of missing material. Hooking up to the power did not go as smoothly as expected, and when it was accomplished the power was insufficient. Worse, the difficulties did not diminish as the work progressed. New ones

86

cropped up. It became clear that—even in the increasingly doubtful event the plant could get into operation—keeping it in operating condition thereafter would probably be impractical. So after most of a year and considerable money had been wasted, Etawah was abandoned and a new site was chosen at Kanpur, a city of some 1,200,000 persons, the largest in Uttar, where industry and commerce had, by Indian standards, been growing rapidly. Space in two unused rooms in an electroplating plant was quickly found. The machinery was installed, the workers hired, and the plant was producing within six weeks. Kanpur possessed not only the space and the electric power, but also repairmen, tools, electricians, bits of needed material, and relatively swift and direct transportation service to other major Indian cities if what was required was not to be found in Kanpur.

I think the little fiasco of Etawah casts light on the great fiasco of Chinese economic planning in 1957-58, so hopefully called The Great Leap Forward. The planners of this program shared with the Rockefellers the belief that village industry would be more wholesome for a predominantly rural country than city industry. In part, the policy seems to have been a defense measure. But it was also, in part, evidently based upon the conventional belief that cities are superficial economically while rural production and rural life are "basic." At any rate, as China was about to launch The Great Leap Forward, the official press agency of the country reported, with alarm, that whereas in 1950 the country had had only five cities with populations of more than a million, it now had thirteen of this size or larger; people inadvisedly kept flocking from the villages to cities where they tended to engage in "unproductive" pursuits, if any. The Great Leap was designed to counter the movement to the cities, as well as to industrialize China rapidly. According to plans, the Chinese economy was to expand at the stupendous rate of forty percent annually by a combination of industrial and agricultural development.

In industry, the growth was to be achieved by building hundreds of thousands of factories each year, scattered, for the most part, among the half million Chinese villages and local market towns. Some of these factories were to produce goods for people in their immediate vicinities. Many were to export to other settlements in their provinces. And some, among them thousands of projected small blast furnaces, were to export to existing industrial centers. Most factories were to be small. In spite of heroic efforts, few of these factories ever got into production; the program was abandoned after two years. The economic corpses of the attempt dot China. The Great Leap was the fiasco of Etawah multiplied by the hundreds of thousands.

"Hospitals and medical organizations in Peking and major provincial cities," says a 1964 Reuters dispatch from Peking, "have formed mobile medical teams to work in the countryside." The measure makes better sense than the policy adopted in The Great Leap Forward when the countryside was expected to originate its own new goods and services. The dispatch continues, "The program, which the Chinese press described as 'a revolutionary measure in our health work' brings specialists to remote village clinics and dispensaries and to larger medical centers in peoples' communes." What we have here is local work developed in cities now being exported from them.

What Is "Basic" Capital?

The orthodox notion that a country's "basic" capital is the land and the labor poured into the land is obviously incorrect. If it were true, then predominantly agricultural countries would nowadays be exporting capital and other financial services to highly industrialized and urbanized countries rather than the reverse. And within industrialized and urbanized countries rural areas would be exporting capital to cities, probably through tax subsidies from country to city. Henry George, reasoning from the premise that land is basic capital and basic wealth, asserted that all profits made in cities derive from the value of city land. Of course the peculiarly high value of city land does not derive from anything inherent in the land, but from the concentrations of work upon city land.

A country's basic wealth is its productive capacity, created by the practical opportunities people have had to add new work to older work.

In the real world, capital originates much as any other city goods do, and rural development is financed by exports of capital from cities.

Discriminatory Use of Capital

People at the bottom of a society customarily find it difficult to get capital for development work. Even if they can get it, they may not be permitted to use it. In societies that are supposedly economically "free," social discrimination and unequal protection of people and their rights by the law (in actual fact, as opposed to theoretical equality of people under law) can effectively prevent many persons from developing their work, no matter what their inherent capacities for doing so may be. In Socialist societies, those who have worked their way up in the bureaucracies have much better access to development capital than others. Whether good use, or any use, is made of such a privilege is another matter.

Considered purely as an economic matter—quite apart from the inherent brutality of social discrimination—the effects of discrimination are not serious in the rural world; an economy can develop in spite of rural caste systems, as happened in medieval Europe and in many other times and places.*

Blacks in the United States have been kept in economic subjection not by their suppression in the rural world where, in any case, they could have added no new jobs to the economy. They have been kept in their economic subjection by discrimination in cities.

*Rural caste systems often disintegrate as economies develop because rural people migrate to cities where they change their social standing. This kind of event was expressed in medieval Europe in the doctrine, "city air makes free," which was applied to serfs who managed to migrate to cities, also, when subsistence farms take on cash crops in developing economies, or farms with cash crops replace manual labor with machines, the old rural social order disintegrates.

When most of the cities in a country neglect their development of new kinds of work, especially by those low in the social hierarchies (whose numbers must either grow in such a situation, or be drained off by emigration to countries with expanding economies), there is nowhere to export the embarrassing superfluity of capital except abroad. The immense exports of capital by the United States during the past quarter century are, in large part, money that was *not* spent in the expensive business of economic trial, error and development by blacks—and others too—in American cities; money that was *not* spent on development of new goods and services to solve acute practical problems in those cities as those problems began to pile up. The embarrassment of riches in an economy that is economizing on development of new work is temporary. It is a prelude to stagnation.

I think, from the symptoms to be observed, that the economy of the United States is in process of stagnating.* Nevertheless, it is still the most advanced economy to be found. Therefore, no matter what its own future may be, it is a suitable economy in which to look for clues to patterns that may be found in more highly developed economies of the future—wherever those economies may prove to be.

The Trend Toward Differentiation

There is a market for standard agricultural tractors and their accessories which are aimed at widespread similarities of needs among farmers, though this is no longer the kind of farm-equipment business that is growing appreciably. As far back as 1961, *Fortune* reported that the giant, mass-production farm-equipment manufacturers were in economic trouble. Their business was static or declining, and

---

*I would not venture to prophesy how decisive this stagnation is. If it proves to be profound and unremitting, it could be comparable to that of the later Roman Empire or to that of many another economy in which revitalization, if it has occurred at all, has followed only upon revolution. If stagnation is still reversible in the United States, then by definition vigorous city-development processes not only can, but will, start into motion again.

they were saddled with huge factories working below capacity and numerous retail outlets that no longer paid their
way. The rapidly growing farm-equipment business was
going disproportionately to more than a thousand small
manufacturers who were aiming precisely at differences
within the market. The big companies had stayed too long
with "the mass concept," *Fortune* commented. "Less of [the
farmers"] equipment money goes for the standard items.
. . . Today a small company can manufacture a highly
specialized item of equipment just as easily as a large firm,
and often at a better profit." Again, the relatively small-
scale *differentiated* equipment production is not a return to
craft methods.

In *Silent Spring,* Rachel Carson attacked the practice of
applying chemical pesticides wholesale—the mass-production approach to pest infestations. Instead, she advocated
differentiated production based upon sophisticated biological controls of varying kinds, according to circumstances.
This is a far cry from depending on the barnyard cat and
the fly swatter, and resigning onself to watching the lo-
custs consume the year's work. It is a far more advanced
approach than indiscriminate, wholesale use of chemicals.
Miss Carson also advocated differentiation of crops within
geographical localities, pointing out that mass production
in farming itself—great factory farms devoted to one kind
of cash crop—leads inherently to drastic imbalances of
natural life and tends to increase the potential ravages of
plant diseases and pests. (It also, I might add, can be economically disastrous to a rural region and often has been.
A rural economy with all its eggs in one basket is bound
to lose out from changes in markets.) As we might expect,
Miss Carson's point has been heeded first in cities. Not
many years ago, for instance, New York City was using the
mass-production approach to street-tree planting. All the
trees planted were London planes which were raised in
great mass-production tree nurseries. As Robert Nichols, a
landscape architect, had been pointing out, some twenty
different varieties of trees do quite as well as London planes
on the city streets; but the city had been committed, under
a powerful administrator, Robert Moses, to mass produc-

tion in this as in all things affecting parks or supervised by the parks department. Now, realizing the wholesale disaster that a London plane tree blight would bring, the city has begun differentiated planting of street trees.

I have brought trees and agricultural equipment into this discussion not only because they illustrate that there is more reason to produce for differences than variations of whims or tastes, but also to show that differentiated production is not a luxury and another term for "custom made." Differentiated production, in spite of its disproportionate requirements for design and development work, is not an extravagance. In real life, real and important differences abound, whether in nature or in a market, whether in the resistance of tree to diseases or in the information about current events needed by people in differing districts. And with economic development all kinds of differentiations increase; they do not diminish.

For some economic needs, mass production is superb. The common denominators are valid and enduring. Mass production is well suited, for example, to brick manufacturing, making screwdrivers, bed sheets, paper, electric light bulbs and telephones. I am not proposing that mass production will disappear from economic life. Farmers still need their standard tractors; people still need standard denim pants or their equivalents. The point is that for some goods, mass production is a makeshift. It represents only an early stage of development and is valid only as an inadequate expedient until more advanced differentiated production has been developed.

With growth of differentiated production in developing economies of the future, we may expect to find other changes in economic life. The average size of manufacturing enterprises will be smaller than at present. But the numbers of manufacturing enterprises will greatly increase and so will the total volume of manufactured goods. Most mass-production enterprises that have not been made obsolete by differentiated production—and many will remain—will have been transplanted to the countryside and into inert towns. There, with their low requirements of labor, their large requirements of space, and their relative self-sufficiency, these industries can operate more efficiently than in cities. Mass-production manufacturing will no longer be regarded as city work. Cities will manufacture even more goods than they do today, but these will be almost wholly differentiated production goods, made in relatively small, or very small, organizations.

Manufacturing work will, I think, no longer be the chief activity around which other economic activities are organized, as it is today and as the work of merchants once was. Instead, services will become the predominant organizational work, the instigators of other economic activities, including manufacturing.

When I was conjecturing how waste recycling systems might be organized in developing economies of the future, I suggested that services would be the key work in such industries, and that the service organizations would be customers for many kinds of waste-collection equipment. This conjecture was based upon the logic of the work, but it corresponds to what I suspect is the coming trend in economic organization generally. Service organizations in developing economies of the future are likely to draw upon products made by many different manufacturers, and are likely to be larger than manufacturing organizations. Even so, they will begin as small businesses and expand as they add innovations.

The primary economic conflict, I think, is between people whose interests are with already well-established economic activities, and those whose interests are with the emergence of new economic activities. This is a conflict that can never be put to rest except by economic stagnation. For the new economic activities of today are the well-established activities of tomorrow which will be threatened in turn by further economic development. In this conflict, other things being equal, the well-established activities and those whose interests are attached to them, must win. They are, by definition, the stronger. The only possible way to keep open the economic opportunities for new activities is for a "third force" to protect their weak and still incipient interests. Only governments can play this economic role. And sometimes, for pitifully brief intervals, they do. But because development subverts the status quo, the status quo soon subverts governments. When development has proceeded for a bit, and has cast up strong new activities, governments come to derive their power from those already well-established interests, and not from still incipient organizations, activities and interests.

In human history, most people in most places most of the time have existed miserably in stagnant economies. Developing economies have been the exceptions, and their histories, as developing economies, have been brief. Now here, now there, a group of cities grows vigorously by the processes I have been describing and then lapses into stagnation for the benefit of people who have already become powerful. I am not one who believes that flying saucers carry creatures from other solar systems who poke curiously into our earthly affairs. But if such beings were to arrive, with their marvelously advanced contrivances, we may be sure we would be agog to learn how their technology worked. The important question however, would be something quite different: What kinds of governments had they invented which had succeeded in keeping open the opportunities for economic and technological development instead of closing them off? Without helpful advice from outer space, this remains one of the most pressing and least regarded problems.

Provided that some groups on earth continue either muddling or revolutionizing themselves into periods of economic development, we can be absolutely sure of a few things about future cities. The cities will not be smaller, simpler or more specialized than cities of today. Rather, they will be more intricate, comprehensive, diversified, and larger than today's, and will have even more complicated jumbles of old and new things than ours do. The bureaucratized, simplified cities, so dear to present-day city planners and urban designers, and familiar also to readers of science fiction and utopian proposals, run counter to the processes of city growth and economic development. Conformity and monotony, even when they are embellished with a froth of novelty, are not attributes of developing and economically vigorous cities. They are attributes of stagnant settlements. To some people, the vision of a future in which life is simpler than it is now and work has become so routine as to be scarcely noticeable, is an exhilarating vision. To other people, it is depressing. But no matter. The vision is irrelevant for developing and influential economies of the future. In highly developed future economies, there will be more kinds of work to do than today, not fewer. And many people in great, growing cities of the future will be engaged in the unroutine business of economic trial and error. They will be faced with acute practical problems which we cannot now imagine. They will add new work to older work.

Bernardo y Adela

Floyd Allen

Chapter 9

These days about the easiest thing to do is to be depressed. Actually, it gets pretty hard sometimes not to be running around in a constant state of dejection. The best that can be said for our daily news is that sometimes it's not as gloomy as it is at other times. Yet, it's not just the daily stream of bad news that's getting to us. What's happening to us is happening to everything we are. Almost every time we take a second look at something, dark clouds of uneasy spirit begin to drift in and surround our thoughts. The other day I was looking through a national real-estate magazine comparing the prices of farms, one state with another, and I was appalled at the low prices for beautiful farms throughout the midwestern states. The magnificent heartland of this country is for sale cheap.

It's not just farm workers moving to the big city in hopes of bigger pay and a good life; farmers and ranchers are leaving too. Sometimes they just pack up and walk away from a 40-year-old dream, and leave behind a profitable operation loaded with conveniences and comfort that wasn't even part of their dreams 40 years ago.

It's been in my mind to make a trip through the Midwest and see if I could develop a better understanding of what it's all about. In the meantime I've been watching farmers and ranchers here in California to see if I could pick up some clues. Can't say for the most part that I have. Just a dark cloud of uneasy spirit.

Take, as an example, a conversation I listened to a while back among a group of California ranchers. These men have been ranching all their lives; as a matter of fact, some of them on ranches established by their grandfathers. They were all sitting around in a large, old-fashioned kitchen in a magnificent—and neglected—ranch house well over a hundred years old. They were drinking hard, talking about hunting. One related as how he'd been watching a dry doe until "she was nice and fat and just right." Another de-

96

scribed a "beautiful shot" he made the year before. All in all, the appearance of the scene and the conversation seemed to be real comradely and friendly, but it wasn't.

It was a nice scene, everybody pretty well off, plenty of 50-dollar boots and 30-dollar hats, clean white dress shirts, nicely weathered blue Levis; separately and all together they made a nice gathering of Marlboro Men. Then I started taking a second look and listening to how things were said, and it really was not friendly at all. There was a subtle and very careful line defining who could say what or claim how much. Couple of the fellows who talked most began to sound more like Cartwright of Bonanza, while three or four of the others were pretty good substitutes of the Virginian.

After a while the conversation did slip, briefly, into the business of ranching. But it mostly centered, with admiration, about how some fellow was making money buying cheap cattle, from down near Mexican way, and his secret was "keep 'em doped up," "anything he thinks they can get, he gives 'em a shot for it." . . . You see, even our memories are depressing, and land is high in California, and farmers are leaving the Midwest. An old American story seems to be coming to an end and a new, not very nice, story seems to have started.

Encounter at the Shopping Center

The other day as I was driving out of the parking lot of our neighborhood shopping center, I happened to glance back and see a Jeep truck that had obviously seen years of hard work. What caught my attention was that it was loaded clear above the cab with empty lug boxes all carefully tied down. Looking closer, I could see a woman sitting in the cab eating an ice-cream cone; and lost to everything else, a man was standing outside—actually he was leaning across the hood of the truck—licking an ice-cream cone too, just staring out across an empty lot at the sunset.

Now you don't see adults—not any more, at least—eat ice-cream cones or anything else with such disinterested contemplation. Inside of me something twanged, something old, something a long time ago. I was almost out on the road when I decided to turn around and go back and talk to them. Now it's somewhat corny and trite to say that gold is where you find it; it is corny but it also happens to be true. I did find gold, and the old American story and, sadly, a clue to the new American story

When I approached them my impression was that they were Japanese, which demonstrates how well I can recognize national characteristics because they are Filipinos. Bernardo is a strongly-built, husky man of fifty-five. He has an agricultural complexion, weathered; you can see the years of wind and sun. When Bernardo is not smiling, the expression upon his face is faintly inquisitive; the rest of the time he's laughing. A soft easy, very easy, pleasing laugh.

Adela is forty-four, small, nicely figured, lighter-complexioned; from a distance you can see that she works in the field; up close, and when she smiles, she is beautiful, desirable, a mother, a wife, Bernardo's sweetheart, his best friend. It would be nice if I could characterize them together in some way that would give you a good picture of them, say like Grant Wood's American Gothic, but that's not their picture. American Romantic perhaps, definitely American Decent. I found in them every virtue that America has described as moral, honest, decent, and worth-while.

At first, when I parked my car next to their truck and got out to talk to them, Bernardo's attitude was cautious, more like non-communicative and Adela pretended to look out of the side window at something on the ground. They stopped eating their ice-cream cones. I felt like an intruder. With Bernardo's poor English and being non-communicative, and my feeling like an intruder, I began wishing that I'd just gone out on the road and gone on home; however I happened to mention ORGANIC GARDENING AND FARMING, Bernardo smiled, "ORGANIC? I know ORGANIC, little magazine, like this?" He described the size of the magazine with his hands, and when I nodded, he nodded also and said, "I got one."

He spoke to Adela, I couldn't understand what he said, and she began pulling things out of the glove compartment until she found two magazines. A *Popular Mechanics* and a 1963 edition of ORGANIC GARDENING AND FARMING. Ten long cents ago. It was 50 cents a copy in 1963.

We talked for a little while after that, cautiously but communicatively. Bernardo helped me to spell his last name correctly and told me how to find his plot. Adela stopped pretending to look out the window and began talking to me about the crops that they raise. She speaks English reasonably well, much better than Bernardo. While we were talking, I noticed that Bernardo was letting his ice cream melt in his cone without eating it, so I asked for permission to come and see them someday while they were working and left.

When I left, it was plain in their faces that they never expected to see me again, and this is the remarkable thing about these people. They are so unassumingly honest that they make no effort to hide their thoughts. You see them, and that's what they are. This kind of honesty can have a strange effect upon you and you start to worry about them, like maybe somebody will try to take advantage of them. After you know them better, you stop worrying. Maybe somebody will try but they're not likely to hurt them very much. Their honesty helps them to arrange their priorities so well that in most things they have a clear picture of what's good for themselves.

I've passed their plot many times from a distance, and from the highway what you see, looking in the direction of their land, is a couple of widely separated, expensive suburban-type homes situated in the middle of nice-looking vegetable crops. Passing by, you have the impression that a couple of farmers have made it pretty good and built themselves a nice home as a reward. When you get off the highway, what you discover is that suburbia—country-style—is moving in, and perhaps in a few short years another chunk of California's best farmland will be gobbled up by houses.

What looked like a large field turned out to be one large field and several small plots, ranging from three to eleven acres. The large field is owned and operated by a Japanese-American family, while the smaller plots are primarily owned and operated by Philippine-Americans. It didn't take much looking around to spot a green Jeep truck parked beside several stacks of lug boxes.

Why Would a Young Man Want to Be a Farmer?

While I was walking toward them I could see that faint inquisitive expression upon Bernardo's face, and it seemed to me that I detected that same non-communicative caution that I had experienced before. But I was mistaken, for as soon as I started talking to them and asking Adela questions, I could see that everything was going to be comfortable and friendly.

When I arrived, Bernardo was loading up the truck, and he was dressed about the same as the first time I saw him. Adela wore a baggy pair of comfortable woolen pants with the bottom of the pantlegs snugly wrapped around her socks and held in place with large rubber bands. She had on a blouse and a green sweater. There wasn't a thing about her that didn't look feminine. She was working on her knees, sorting and boxing Italian squash. All the while I was there she kept working, pausing from time to time to think of something, or looking up to tell me something, or looking over at Bernardo to laugh with him—but her hands were always moving.

The first thing you notice right away with Bernardo and Adela is that they minimize themselves. They talk about their work and what they do like it's not much at all; and if you're careful, you can get the impression that they just do one unimportant thing, and then the next one unimportant thing until it's all done. Oh, they admit that they do do a lot of work, lots of it, but they treat it like it's really nothing much at all. At one point during our conversation I asked them what advice would they give a young man just starting out to be a farmer. The question puzzled them. I pressed them for some kind of an answer, and what they finally said approximated a question for *me*. Why would a young man born in this country and having an

100

education want to be a farmer and do the kind of work they do?

They have two children, a girl, fifteen, and a boy who is fourteen. Timidly, almost as though they'd rather not speak about it out loud, they reveal that they have expectations for their children, a hope that their children will be—another old, American story—they want their kids to be better than they are. Even now, they're a little in awe of the possibilities.

Taking the line of questioning along this direction, I asked them if their son helps them out on weekends. (They work seven days a week.) Yes, he's a good boy and gives them lots of help, often coming out on weekends to do the irrigating and sometimes he drives the tractor. I asked Adela if her daughter comes down and works on weekends. "No!" She looked up at me and then kind of looked across the field. "She stays home . . . helps with housework . . . helps with cooking." I waited a few minutes and then asked her, "Adela, would you like your son to become a farmer?" "No!" That was the loudest she ever spoke to me. They looked at each other and then she continued, "want them to go to school . . . that's what we tell them all the time. Go to school, get good education and not be like us and work all the time."

They do work long hours, and frequently long, hard hours. Adela gets up early to do her housework, and then puts in an average ten-hour day in the field before going home to prepare dinner for the family. Bernardo continues to work on for a few more hours, and then goes home when dinner is ready. They work seven days a week, starting in the spring and until harvest is over. No days off. No Sundays. No holidays. No vacations. "Take vacation in wintertime when there's no work to do."

In this part of California wintertime means the rainy season, and vacation to them means driving a few miles to the beach and taking along a picnic lunch. Bernardo will probably do a little fishing. It's also a time for visiting and being visited, and good conversation. They have warm friends. While I was talking to them, a friend stopped by and I could feel the presence of their mutual respect. The kind of feeling that people have for each other after long years of sharing experiences and achieving similar goals. I couldn't help but notice that their friend has the same easy, pleasing laugh that bubbles so spontaneously from Bernardo and Adela.

How Vast the Gulf of Obstacles

Putting everything that was said together and trying to make allowances for a certain degree of language confusion and the fact that I was a stranger asking personal questions, the conversation left me feeling a little uncertain if not just plain bewildered. Who knows how vast the gulf separating the obstacles that those two people had to overcome to be what they are today? Perhaps they have literally clawed down mountains. Yet, they keep putting themselves down and making their own efforts as nothing. Without a doubt they're giving their children the idea that to be better also means to do a different kind of work.

On a small, a very small three-acre plot of land, those two people working together manage to average as much as six thousand dollars a year gross. Working that land they have comfortably fed and clothed themselves and their children. They own a comfortable home which they built themselves, working on it little by little each winter. As near as I could determine, they don't owe a cent. Their equipment is top-notch and kept in good working condition. They have a bank account. They pay taxes. They are an asset to their community, and by their efforts they have contributed to the food on our tables. Still they keep talking themselves down and they want their children to do better. Do better what?

I'm sure that they would like to work a little less than they do, that they would like to take off a little once in a while, perhaps one day a week and spend more time being with their children and visiting with friends; but something sticks in my mind. They never even once came near to implying that they didn't *like* the work they do, or that *they* actually minded the long hours. It's just that they think that someone smarter—better—wouldn't do it.

102

Running my memories backwards, back about 15 years ago when I was traveling through the Midwest, I recalled various conversations I had with a number of farmers. I tried to remember my feelings; then I remembered. I was disturbed then, bothered, disappointed. The conversations were mostly sour grapes. When I tried to talk about farming—living on and operating a farm—with those farmers, the conversation always drifted around to fine new pieces of expensive equipment or to a profitable crop. In other words, their point of pride always got around to money. And the thing is, that not all of those men were all that much in love with money.

Looking back at them and thinking about Bernardo and Adela, I can detect similar attitudes. "Want you to stay in school and go to college, get a good education and make something worth-while out of yourself." The same old story, and the undercurrent running under everything they taught their kids. "A really smart man, if he was really smart wouldn't do this kind of work for a living." There you have it. They talked about money and a good life and they had money and a good life. So what were they talking about, what did they mean, "smart man"?

"Talking Themselves and Their Kids Right Out of the Country"

Even in those days you could drive along the highways of the Midwest and see beautiful well-built farmhouses that had been abandoned. The farmer had probably moved to town, had become a "sidewalk farmer," and the old place didn't mean a thing any more. The good life had moved elsewhere. Right about here a person can get all mixed up talking about changing values and lot of abstractions that won't tell anything. The clue to the Midwest is the clue to Bernardo and Adela. "Better" means "beautiful people"; "good life" means doing what the beautiful people do. Farming is out; maybe ranching, especially big ranching is in, but raising crops and milking cows is out.

By and large the kids have, and are, taking their parents' advice. They're leaving the farms to be better, out looking for "worth-while." Their parents taught them to do that

103

and now they're leaving, leaving farming, leaving farms, leaving "insignificant." And from the way it looks in the catalogs, the parents are leaving too. Now I don't say that one clue is a whole explanation, but I do say that farmers are and have been talking themselves and their kids right out of the country, and the fresh air and a private contemplative way of life. And will somebody tell me what is necessarily better or worth-while doing one thing that might be needed over another thing that is needed?

Where Cities and Farms Come Together

Wendell Berry

The mentality of organic agriculture is not a technological mentality—though it concerns itself with technology. It does not merely ask what is the easiest and cheapest and quickest way to reach an immediate aim. It is, rather, a complex and radical attitude toward the problem of our relation to the earth. It is concerned with the long-term questions of what humans need from the earth, and what duties and devotions humans owe the earth in return for the satisfaction of their needs. It understands that the terms of a lasting agriculture are not human terms, that the final terms are nature's, that an agriculture—and for that matter, a culture—that holds in ignorance or contempt the truths and the mysteries of nature is doomed to failure, for it is out of control.

Chapter 10

At least since the time of Henry Adams, numerous critics and historians have been concerned with the disintegration of the synthesis of disciplines that made the medieval cathedral one of the supreme articulations of humanity's relation to God. Only recently have we begun to be aware of the disintegration of an even more ancient and fundamental synthesis—that of the old peasant and yeoman agriculture, which still stands as the best articulation of humanity's relation to the world. This was not simply an agriculture; at best, it was also a *culture* of such deep-rooted and complex wisdom that it preserved the fertility of the earth under the most intensive human use. It was a culture that made men the preservers rather than the parasites of the sources of their life. The organic movement has its roots in this ancient agriculture that was so wise and profound a bond between human beings and their fields. And it is the rise of the organic movement that affords us a perspective from which we can understand the consequences of the disintegration of that bond—a disintegration that now palpably threatens the destruction, not merely of human culture, but of human life as well.

Nearly all the old standards, which implied and required rigorous disciplines, have now been replaced by a new standard of efficiency, which requires, not discipline, not a mastery of means, but rather a carelessness of means, a relentless subjection of means to immediate ends. The standard of efficiency displaces and destroys the standards of quality because, by definition, it cannot even consider them. Instead of asking a man what he can do well, it asks him what he can do fast and cheap. Instead of asking the farmer to practice the best husbandry, to be a good steward and trustee of his land and his art, it puts irresistible pressures on him to produce more and more food and fiber more and more cheaply, thereby destroying the health of the land, the best traditions of husbandry, and the farm population itself. And so when we examine the principle of efficiency as we now practice it we see that it is not really efficient at all. As we use the word, efficiency means no such thing, or it means short-term or temporary efficiency, which is a contradiction in terms. It means cheapness at any price. It means hurrying to nowhere. It means the profligate waste of humanity and of nature. It means the greatest profit to the greatest liar. What we have called efficiency has produced among us, and to our incalculable cost, such unprecedented monuments of destructiveness and waste as the strip mining industry, the Pentagon, the federal bureaucracy, and family car.

Real efficiency is something entirely different. It is neither cheap (in terms of skill and labor) nor fast. Real efficiency is long-term efficiency. It is to be found in means that are in keeping with and preserving of their ends, in methods of production that preserve the sources of production, in workmanship that is durable and of high quality. In this age of consumerism, planned obsolescence, frivolous horsepower and surplus manpower, those salesmen and politicians who talk about efficiency are talking, in reality, about spiritual and biological death.

Specialization, a result of our nearly exclusive concern with the form of exploitation that we call efficiency, has in its turn become a destructive force. Carried to the extent to which we have carried it, it is both socially and ecologically destructive. That specialization has vastly increased our knowledge, as its defenders claim, cannot be disputed. But I think that one might reasonably dispute the underlying assumption that knowledge per se, undisciplined knowledge, is good. For while specialization has increased knowledge, it has fragmented it. And this fragmentation of knowledge has been accompanied by a fragmentation of discipline. That is, specialization has tended to draw the specialist toward the discipline that will lead to the discovery of new facts or processes within a narrowly defined area, and it has tended to lead him away from or distract him from those disciplines by which he might consider the *effects* of his discovery upon human society or upon the world. It has tended to value the disciplines that pertain to the gathering of knowledge and to its immediate use, and to devalue those that pertain to its ultimate effects.

Nowhere are these tendencies more apparent than in agriculture. For years now the agricultural specialists have tended to think and work in terms of piecemeal solutions and in terms of annual production, rather than in terms of a whole and coherent system that would maintain the fertility and the ecological health of the land over a period of centuries. Focused nearly exclusively upon so-called efficiency with respect to production, as if the only discipline pertinent to agriculture were that of economics, they have eagerly abetted a rapid industrialization of agriculture which is potentially catastrophic, both in the ecological deterioration of farm areas and in the diminishment, the dispossession and displacement, of the rural population.

Ignoring the ample evidence that a healthy agriculture is highly diversified, using the greatest possible variety of animals and plants, and that it returns all organic wastes to the soil, the specialists of the laboratories have promoted the specialization of the farms, encouraging one-crop agriculture and the replacement of humus by chemicals. And as the pressures of urban populations upon the land have grown, the specialists have turned more and more, not to the land, but to the laboratory.

Ignoring the considerable historical evidence that to have a productive agriculture over a long period of time it is necessary to have a stable and prosperous rural population closely bound in sympathy and association to the land, the specialists have either connived in the dispossession of small farmers by machinery and technology, or have actively encouraged their migration into the cities.

The result of the short-term vision of these experts is a whole series of difficulties that together amount to a rapidly building ecological and social disaster, which there is little disposition at present to regret, much less to correct. The organic wastes of our society, for which our land is starved, and which in a sound agricultural economy would be returned to the land, are instead flushed out through the sewers to pollute the streams and rivers and, finally, the oceans; or they are burned and the smoke pollutes the air; or they are wasted in other ways. Similarly, the small farmers who in a healthy society would be the mainstay of the country—whose allegiance to their land, continuing and deepening in association from one generation to another, would be the motive and guarantee of good care—are forced out by the homicidal economics of efficiency, to become emigrants and dependents in the already overcrowded cities. In both instances, by the abuse of knowledge in the name of efficiency, assets have been converted into problems.

Modern agricultural practice concentrates almost exclusively on the productive phase of the natural cycle. The means of production become more elaborate all the time, but the means of return—the building of health and fertility in the soil—are reduced more and more to the shorthand of chemicals. According to the industrial vision of it, the life of the farm does not rise and fall in the turning cycle of the year; it goes on in a straight line from one harvest to another. In the long run this may well be more productive of waste than of anything else. It wastes the soil. It wastes the animal manures and other organic residues that industrialized agriculture fails to return to the soil. And what may be our largest agriculture waste is not usually recognized as such, but is thought to be both an urban product and an urban problem: the tons of garbage and sewage that are burned or buried or flushed into the rivers. This, like all waste, is the abuse of a resource. It was ecological stupidity of exactly this kind that destroyed Rome. The chemist Liebig wrote that "The sewers of the immense metropolis engulfed in the course of centuries the prosperity of Roman peasants. The Roman Campagna would no longer yield the means of feeding her population; these same sewers devoured the wealth of Sicily, Sardinia and the fertile lands of the coast of Africa."

To recognize the extent and the destructiveness of our "urban waste" is to recognize the shallowness of the notion that agriculture is only another form of technology to be turned over to a few specialists. The sewage and garbage problem of our cities suggests, rather, that a healthy agriculture is a cultural organism, not merely a universal necessity but a universal obligation as well. It suggests that, just as the cities exist within the ecology, they also exist within agriculture. It suggests, that, like farmers, city-dwellers have agricultural responsibilities: to use no more than necessary, to waste nothing, to return organic residues to the soil.

We are being virtually buried by the evidence that those disciplines by which we manipulate *things* are inadequate disciplines. Our cities have become almost unlivable because they have been built to be factories and vending machines rather than communities. They are conceptions of the desires for wealth, excitement, and ease—all illegitimate motives from the standpoint of community, as is proved by the fact that without the community disciplines that make for a stable, neighborly population the cities have become scenes of poverty, boredom, and disease.

The rural community—that is, the land and the people— is being degraded in complementary fashion by the specialists' premise that the exclusive function of the farmer is production and that his major discipline is economics. On the contrary, both the function and the discipline of the farmer have to do with provision: he must provide, he must look ahead. He must look ahead, however, not in the economic-mechanistic sense of anticipating a need and fulfilling it, but in the sense of using methods that preserve the source. In his work sound economics becomes identical with sound ecology. The farmer is not a factory worker, he is the trustee of the life of the topsoil, the keeper of the rural community. In precisely the same way the dweller in a healthy city is not an office or a factory worker, but part and preserver of the urban community. It is in thinking of the whole citizenry as factory workers —as readily interchangeable parts of an entirely mechanistic and economic order—that we have reduced our people to the most abject and aimless of nomads, and displaced and fragmented our communities.

An index of the health of a rural community—and, of course, of the urban community, its blood kin—might be found in the relative acreages of field crops and tree crops. By tree crops I mean not just those orchard trees of comparatively early bearing and short life, but also the fruit and nut and timber trees that bear late and live long. It is characteristic of an unsettled and anxious farm population —a population that feels itself, because of economic threat or the degradation of cultural value, to be ephemeral—that it farms almost exclusively with field crops, within economic and biological cycles that are complete in one year. This has been the dominant pattern of American agriculture. Stable, settled populations, assured both of an economic sufficiency in return for their work and of the cultural value of their work, tend to have methods and attitudes of a much longer range. Though they have generally also farmed with field crops, established farm populations have always been planters of trees. In parts of Europe, according to J. Russell Smith's important book, *Tree Crops,* the steep hillsides were covered with orchards of chestnut trees which were kept and maintained with great care by the farmers. Many of the trees were ancient, and when one began to show signs of dying a seedling would be planted beside it to replace it. Here is an agricultural discipline that could only develop among farmers who felt secure—as individuals, and also as families and communities—in their connection to their land. Such a discipline depends not just on the younger men in the prime of their workdays but also on the old men, the keepers of tradition. The model figure of this agriculture is an old man planting a young tree that will live longer than a man, and that he himself may not live to see in its first bearing. And he is planting, moreover, a tree whose worth lies beyond any conceivable market prediction. He is planting it because the good sense of doing so has been clear to men of his place and kind for generations. The practice has been continued because it is ecologically and agriculturally sound; the economic soundness of it must be assumed. While the planting of a field crop, then, may be looked upon as a "short-term investment," the planting of a chestnut tree is a covenant of faith.

The metaphor governing the distortions of efficiency and specialization has been that of the laboratory. The working assumption has been that nature and society, like laboratory experiments, can be manipulated by processes that are for the most part comprehensible toward ends that are for the most part foreseeable. But the analogy, as any farmer would know instantly, is too simple, for both nature and humanity are vast in possibility, unpredictable, and ultimately mysterious. Sir Albert Howard was speaking to this problem in *An Agricultural Testament:* "Instead of breaking up the subject into fragments and studying agriculture in piecemeal fashion by the analytical methods of science, appropriate only to the discovery of new facts, we must adopt a synthetic approach and look at the wheel of life as one great subject and not as if it were a patchwork of unrelated things." A much more appropriate model for the agriculturist, scientist or farmer, is the forest, for the forest, as Howard pointed out, "manures itself" and is therefore self-renewing; it has achieved that "correct relation between the processes of growth and the processes of decay that is the first principle of successful agriculture." A healthy agriculture can take place only within nature, and in cooperation with its processes, not in spite of it and not by "conquering" it. Nature, Howard points out, in elaboration of his metaphor, "never attempts to farm without live stock; she always raises mixed crops; great pains are taken to preserve the soil and to prevent erosion; the mixed vegetable

111

and animal wastes are converted into humus; *there is no waste* [my emphasis]; the processes of growth and the processes of decay balance one another; ample provision is made to maintain large reserves of fertility; the greatest care is taken to store the rainfall; both plants and animals are left to protect themselves against disease."

The fact is that farming is not a laboratory science, but a science of practice. It would be, I think, a good deal more accurate to call it an art, for it grows not only out of factual knowledge but out of cultural tradition; it is learned not only by precept but by example, by apprenticeship; and it requires not merely a competent knowledge of its facts and processes, but also a complex set of attitudes, a certain culturally evolved stance, in the face of the unexpected and the unknown. That is to say that it requires *style* in the highest and richest sense of that term.

One of the most often repeated tenets of contemporary optimism asserts that "a nation that can put men on the moon certainly should be able to solve the problem of hunger." This proposition seems to me to have three important flaws, which I think may be taken as typical of our official view of ourselves:

1. It construes the flight to the moon as an historical event of a complete and coherent significance, when in fact it is a fragmentary event of very uncertain significance. Americans have gone to the moon as they came to the frontiers of the New World: with their minds very much upon getting there, very little upon what might be involved in staying there. I mean that because of our history of waste and destrutcion here, we have no assurance that we can survive in America, much less on the moon. And until we can bring into balance the processes of growth and decay, the white man's settlement of this continent will remain an incomplete event. When a Japanese peasant went to the fields of his tiny farm in the pre-industrial age, he worked in the governance of an agricultural tradition thousands of years old, which had sustained the land in prime fertility during all that time, in spite of the pressures of a population that in 1907 had reached a density, according to F. H. King's *Farmers of Forty Centuries,* of "more than three

people to each acre." Such a farmer might look upon his crop year as a complete and coherent historical event, suffused and illuminated with a meaning and mystery that were both its own and the world's, because in his mind and work agricultural process had come into an enduring and preserving harmony with natural process. To him the past confidently promised a future. What are we to say, by contrast, of a society that places no value at all upon such a tradition or such a man, that instead works the destruction of such imperfect agricultural traditions as it has, that replaces the farm people with machines, that values the techniques of production far above the techniques of land maintenance, and that has espoused as an ideal a depopulated countryside farmed by a few technicians for the supposedly greater benefit of hundreds of millions crowded into cities and helpless to produce food or any other essential for themselves?

2. The agricultural optimism that bases itself upon the moon landings assumes that there is an equation between agriculture and technology, or that agriculture is a kind of technology. This grows out of the much-popularized false assumptions of the agricultural specialists, who have gone about their work as if agriculture was answerable only to the demands of economics, not to those of ecology or human culture, just as most urban consumers conceive eating to be an activity associated with economics but not with agriculture. The ground of agricultural thinking is so narrowly circumscribed, one imagines, to fit the demands of laboratory science, as well as the popular prejudice that prefers false certainties to honest doubts. The discipline proper to eating, of course, is not economics but agriculture. The discipline proper to agriculture, which survives not just by production but also by the return of wastes to the ground, is not economics but ecology. And ecology may well find its proper disciplines in the arts, whose function is to refine and enliven *perception,* for ecological principle, however publically approved, can be enacted only upon the basis of each man's perception of his relation to the world.

Under the governance of the laboratory analogy, the *device,* which is simple and apparently simplifying, becomes the focal point and the standard rather than the human need, which is complex. Thus an agricultural specialist, prescribing the best conditions for the use of a harvesting machine, thinks only of the machine, not its cultural or ecological effects. And because of the fixation on optimum conditions big-farm technology has come to be highly developed, whereas the technology of the family farm, which must still involve methods and economies that are "old fashioned," has been neglected. For this reason, and others perhaps more pressing, small-farm technology is rapidly passing from sight, along with the small farmers. As a result we have an increasing acreage of supposedly "marginal," but potentially productive, land for the use of which we have neither methods nor people—an alarming condition in view of the likelihood that someday we will desperately need to farm these lands again.

The drastic and incalculably dangerous assumption is that farming can be considered apart from farmers, that the land may be conceptually divided in its use from human need and human care. The assumption is that moving a farmer into a factory is as simple a cultural act as moving a worker from one factory to another. It is inconceivably more complicated, and more final. American agricultural tradition has been for the most part inadequate from the beginning, and we have an abundance of diminished land to show for it. But American farmers are nevertheless an agricultural population of long standing. Most settlers who farmed in America farmed in Europe. The farm population in this country therefore embodies a knowledge and a set of attitudes and interests that have been literally thousands of years in the making. This mentality is, or was, a great resource upon which we might have built a truly indigenous agriculture, fully adequate to the needs and demands of American regions. Ancient as it is, it is destroyed in a generation in every family that is forced off the farm into the city—or in less than a generation, for the farm mentality can survive only in sustained vital contact with the land.

A truer agricultural vision would look upon farming not as a function of the economy or even of the society, but as a function of the land; and it would look upon the farm population as an indispensable and inalienable part of the ecological system. Among the Incas, according to John Collier, the basic social and economic unit was the tribe or *ayllu,* but he says "the *ayllu* was not merely its people, and not merely the land, but people and land wedded through a mystical bond." The union of the land and the people was indissoluble, like marriage or grace. Chief Rekayi of the Tangwena tribe of Rhodesia, in refusing to leave his ancestral home which had been claimed by the whites, is reported in a recent newspaper account to have said: "I am married to this land. I was put here by God . . . and if I am to leave, I must be removed by God who put me here." This altogether natural and noble sentiment was said by the Internal Affairs Minister to have been "Communist inspired."

3. The notion that the moon voyages provide us assurance of enough to eat exposes the shallowness of our intellectual confidence, for it is based upon our growing inability to distinguish between training and education. The fact is that a man can be made an astronaut much more quickly than he can be made a good farmer, for the astronaut is produced by training and the farmer by education. Training is a process of conditioning, an orderly and highly efficient procedure by which a man learns a prescribed pattern of facts and functions. Education, on the other hand, is an obscure process by which a person's experience is brought into contact with his place and his history. A college can train a person in four years; it can barely begin his education in that time. A person's education begins before his birth in the making of the disciplines, traditions and attitudes of mind that he will inherit, and it continues until his death under the slow, expensive, uneasy tutelage of his experience. The process that produces astronauts may produce soldiers and factory workers and clerks; it will never produce good farmers or good artists or good citizens or good parents.

White American tradition, so far as I know, contains only one coherent social vision that takes such matters into consideration, and that is Thomas Jefferson's. Jefferson's public reputation seems to have dwindled to that of Founding Father and advocate of liberty, author of several documents and actions that have been enshrined and forgotten. But in his thinking democracy was not an ideal that stood alone. He saw that it would have to be secured by vigorous disciplines, or its public offices would become merely the hunting grounds of mediocrity and venality. And so those who associate his name only with his political utterances miss both the breadth and depth of his wisdom. As Jefferson saw it, two disciplines were indispensable to democracy: on the one hand education, which was to produce a class of qualified leaders, an aristocracy of "virture and talents" drawn from all economic classes; and on the other hand, land, the widespread possession of which would assure stable communities, a tangible connection to the country, and a permanent interest in its welfare. In language which recalls Collier's description of the *Ayllu* of the Incas, and the language of Chief Rekayi of the Tangwenans, he wrote that farmers "are tied to their country, and wedded to its liberty and interests, by the most lasting bonds." And: ". . . legislators cannot invent too many devices for subdividing property . . ." And: ". . . it is not too soon to provide by every possible means that as few as possible shall be without a little portion of land. The small landholders are the most precious part of a state . . ." For the discipline of education of the broad and humane sort that Jefferson had in mind, to produce a "natural aristocracy . . . for the instruction, the trusts, and government of society," we have tended more and more to substitute the specialized training that will most readily secure the careerist in his career. For the ownership of "a little portion of land" we have, and we apparently wish, to substitute the barbarous abstraction of nationalism, which puts our minds within the control of whatever demagogue can soonest rouse us to self-righteousness.

On September 10, 1814, Jefferson wrote to Dr. Thomas Cooper of the "condition of society" as he saw it at that time: ". . . we have no paupers, the old and crippled among us, who possess nothing and have no families to take care of them, being too few to merit notice as a separate section of society . . . The great mass of our population is of laborers; our rich . . . being few, and of moderate wealth. Most of the laboring class possess property, cultivate their own lands . . . and from the demand for their labor are enabled . . . to be fed abundantly, clothed above mere decency, to labor moderately. . . . The wealthy . . . know nothing of what the Europeans call luxury." This has an obvious kinship with the Confucian formula: ". . . that the producers be many and that the mere consumers be few; that the artisan mass be energetic and the consumers temperate . . ."

In the loss of that vision, or of such a vision, and in the abandonment of that possibility, we have created a society characterized by degrading urban poverty and an equally degrading affluence—a society of undisciplined abundance, which is to say a society of waste.

INDEX